Principles and Practice of Agricultural Meteorology

Principles and Practice of Agricultural Meteorology

S. Venkataraman

Retired Director, India Meteorological Department

and

Former United Nations (WMO/FAO) Agromet. Expert

BSP **BS Publications**

A unit of **BSP Books Pvt. Ltd.**

4-4-309/316, Giriraj Lane, Sultan Bazar,
Hyderabad - 500 095

Phone : 040 - 23445605, 23445688

Principles and Practice of Agricultural Meteorology *by S. Venkataraman*

© 2015 *by Publisher*

Published by

BSP **BS Publications**

A unit of **BSP Books Pvt. Ltd.**

4-4-309/316, Giriraj Lane, Sultan Bazar,

Hyderabad - 500 095

Phone : 040 - 23445605, 23445688

e-mail : info@bspbooks.net

ISBN: 978-93-85433-98-6 (HB)

Foreword

I am happy that Dr. S. Venkataraman has written a detailed book on the management of weather responses of crops as well as the impact of climate change. Meteorological factors play a central role in determining the output of crops. Agriculture is applied ecology and therefore the science of agro-ecology needs much greater attention than it is receiving at present. This book corrects this deficiency and enunciates principles governing the interaction between environment and agriculture.

Farming is very much influenced by three major climatic factors. First, the date of onset of soaking rains determines the choice of the date for planting the new crop. Second, the interspell duration of crops is exceedingly important from the point of view of productivity and stability. Third, the moisture holding capacity of the soil is exceedingly important. It will determine the fate of the crop when the interspell duration between two rains exceeds about ten days. Dr. Venkataraman's timely book will help us to undertake monsoon management in a scientific manner. I would therefore like to discuss some aspects of monsoon management in relation to crop production, in order to emphasise the importance and timeliness of this book.

Advance forecasts by the South Asian Climate Outlook Forum and the India Meteorological Department indicate that the South West Monsoon rainfall may be deficient this year. Also, there is a possibility of evolution of an El Nino event during June to September. There is a 45% probability that central, west, north-west and south India will get below normal rainfall. There is also a 40% chance that eastern states like Odisha, West Bengal, North east and most of Jammu and Kashmir may get normal rains during the S W monsoon period.

In India, unlike in USA and Australia, agriculture is not just a food producing enterprise but is the backbone of the livelihood security of over 60% of the population. Therefore there is no time to relax in the area of taking anticipatory steps to safeguard food, water, energy and livelihood security in rural India, in the event of erratic monsoon. The recent report of the Inter-Government Panel on Climate Change has warned that we will face both higher mean temperatures and a rise in sea level, if we do not take action on cutting down green house gas emissions. The low carbon pathway of development still remains a topic for academic discussion rather than for political and practical action. International prices of food and other agricultural commodities tend to remain volatile. This is why I have frequently emphasized that the future belongs to nations with grains and not guns. The right to food enshrined in the National Food Security Act can be implemented only with the help of our farmers, unlike the right to information which can be implemented with the help of files. Right now the Government has enough stock to fulfill the legal obligation to provide 5 kgs of wheat, rice or millets per month to nearly 75% of our population. With one widespread drought, the current food stocks may disappear.

Fortunately, we still have a large untapped production reservoir in most food grains even with technologies available on the shelf. What is important is the mobilization of group endeavour among farm families with small holdings in areas such as plant protection, water harvesting and post-harvest technologies. Water harvesting in homes, farms and factories must become mandatory. Tamil Nadu has already initiated steps in this direction. The rain cum solar energy centre functioning in Chennai is a source of credible public information on rain water harvesting and solar energy use. Such Centres need to be replicated in all our cities, towns and block headquarters.

The latest technologies for using the available water in the most efficient manner should be adopted. As regards fertilizer use efficiency, there are technologies such as those developed by the International Fertilizer Development Centre in USA which can help to improve the efficiency of urea use by about 50%. Methods of managing the triple alliance of pests, pathogens and weeds must be popularized.

Besides causing food and water shortage, deficient rainfall adds to the problem of energy shortage. Every calamity also provides an opportunity for progress. Harvesting of sun in homes, offices, fields and factories should also become mandatory, since this can make an important contribution to increasing energy supply both in rural and urban areas. At the post-harvest stage, it is important that a National Grain Storage Policy consisting of the following three components is adopted. First, we should promote the use of small storage bins like the Pusa Bin at the farm level. Second, we should implement the rural godown scheme for safe storage of both food grains and perishable commodities at the village level. Such a rural godown scheme was introduced as early as 1979, but the programme is yet to take off in a manner that can make a difference to preventing the loss of food items at the village/block level. Third, we should establish a National Grid of Ultramodern Silos at least at 50 locations in the country, each capable of storing about a million tonnes of food grains. Unfortunately there is currently a serious mismatch between production and post-harvest technologies, with the result that both producers and consumers do not get the full benefit of higher production.

For the food and water security of farm animals, we need to earmark potential areas for establishing cattle camps where the animals can be looked after during a drought emergency. Such cattle camps should have access to water. A suggestion I had made over three decades ago that we should identify and establish Ground Water Sanctuaries at appropriate places is yet to be implemented. Such Ground Water Sanctuaries are concealed aquifers which should be tapped only when absolutely essential. Like a Wild Life Sanctuary, they should be protected from exploitation. The establishment of Ground Water Sanctuaries at least one each in the 130 agro-climatic zones generally identified in our country, will help us to save precious cattle and other farm animals, both from distress sale and starvation deaths.

In my Sardar Patel Lectures of the All India Radio delivered in 1973, I had suggested that we should develop Drought, Flood and Good Weather Codes both to minimize the adverse impact of unfavourable monsoons and to maximize production in a good monsoon year. The Drought Code consists of a series of do's and don'ts during deficient rainfall. Both in the case of Drought and Flood Codes, Seed Banks consisting of seeds of alternative crops should be maintained. Seed reserves are as important for crop security as grain reserves are important for food security. For example during the recent severe drought in California, it was found that some of the earlier crops like millets survived and gave reasonable yield, while wheat or rice could not withstand the severe drought.

Fortunately in our National Food Security Act, there is provision to procure and supply under the Public Distribution System, local grains like ragi, bajra, jowar and a whole series of minor millets.

Since such crops require milling, they have been referred to as Coarse Cereals. They should be really referred to as Climate Smart Nutri Cereals. Such underutilized crops are rich in many macro and micro-nutrients and could help to fight both protein hunger caused by the deficiency of protein in the diets and hidden hunger caused by the deficiency of micronutrients such as iron, iodine, zinc, Vitamin A and Vitamin B 12. If the Food Security Act is backed up with a Nutrition Literacy Movement, the demand for Climate Smart Nutri-cereals will grow. This will help to promote the cultivation of crops which may do better under drought conditions.

In each of the major agro-climatic zones, at least one woman and one male member of every panchayat or local body should be trained as Climate Risk Manager. They can help the rest of the community in implementing the provisions of the proposed Drought, Flood and Good Weather Codes. The Government has introduced recently a National Policy for Agroforestry. Agroforestry combines the benefits of carbon sequestration and local food security. The inclusion of fertilizer trees in agro-forestry systems can help to build soil carbon banks.

2014 is the International Year of Family Farming. Family farming is both a way of life and a means to livelihood. India has probably the largest number of family farmers. Our aim should be to make Every Family Farm a Climate Smart Farm, equipped with the knowledge and technologies essential to manage the expected El Nino triggered adverse rainfall conditions.

I hope that the principles and methods enunciated in this book will get incorporated in the manuals relating to agromet aspects of irrigation scheduling, climate and rainfall zonations and other allied events. We owe a deep sense of gratitude to Dr. S. Venkataraman for this labour of love in the cause of climate resilient agriculture. I hope this book will be widely read and used both by professionals and policy makers, since it is based on rich practical experience and professional knowledge.

Dr. M. S. Swaminathan

Retired Director General Indian Council of
Agricultural Research and Founder-Chairman,
M.S. Swaminathan Research Foundation,
Chennai.

Preface

The type and fertility status of soil, which have a bearing on weather-responses of crops, carry a climatic imprint. Environmental conditions prevailing during maturity, storage and transport affect quality of seeds and planting material and their post-sowing/planting behaviour. Thus, climatic and weather factors are operative even before a crop is sown. Weather factors exert (i) a direct effect on crops through the provision or denial of congenial conditions for their optimal growth, development and yield and (ii) indirect effects through (a) assistance to or interference with cultural operations (b) outbreak of pests and diseases (c) physical damage and (d) erosion of top soil. Thus, there is no aspect of crop culture that is free of the impact, direct or indirect, of weather factors. Weather factors also determine the water needs and fertilizer requirements of crops and ease of availability of soil moisture.

The science of Agroecology has, however, not received the needed attention, as neglect of ecological considerations in crop production leads to widespread but slow degradation of soil, air and water environments in an imperceptible way in rural areas. Again, in the management of farmlands, compared to soil and its relation to macro and micro biota, less attention has been paid as to what climatic factors affect the agricultural ecosystems and how they do so. Thus, neglect of weather influences on production and protection of crops must, a priori, lead to non-sustainability of crop production as is being witnessed currently. By the same token agroclimatic analyses should help in sustainability of crop production. However, agrometeorological indices and zonations can only be comparative in nature. Even for comparative purposes, the agrometeorological methodology, to arrive at the output, needs to be rationalized and standardized so that the end product of an agromet analyses, validated with ground truth at a few centers, can be extrapolated in time and space through meteorological links.

In view of the above, in this book, the emphasis is on detailing the principles governing weather relations of (a) production, protection, management, monitoring and assessment of crops and (b) use-efficiencies of resources in production and protection of crops. Agrometeorology should deal with both annual and perennial crops.

Unraveling the weather relations of biotic and non-biotic stresses on yields and use of climate data and weather information for betterment of production and protection of perennial crops is a highly daunting task. In this book, attention has, therefore, been confined to the agrometeorology of annual crops. The contents have been organised to develop the subject logically as detailed below.

In Chapter 1, important pioneering work prior to 1930, when agrometeorology came to be recognised as a specialised science, like quantification of crop-weather relationship as early as 1753, pathways in formation of photosynthates, assessment of weather requirements of crops, need for understanding role of weather in crop diseases and biological control of pests, influence of windbreaks on crop yields, weather influences on crop water needs, weather aspects of incidence of crop pests, photoperiodism and recognition of soil temperature as an ecological factor are mentioned. Developments in or relating to agrometeorology are given decade-wise from the 1930s to 1990s and in the first decade of the current century.

Chapter 2 deals with (a) the self-created micro-climates of crops and natural vegetation and their agricultural significance (b) role of crop morphology in radiation and water use-efficiencies and in emission of Methane into the air from puddled rice (c) influence of growth stages on (i) susceptibility of crops to abiotic and biotic stresses and (ii) reception of photo and thermal stimuli for phasic development (d) physiological reasons for differences amongst crops in (i) response to a given change in environment on their growth, development and yield (ii) water requirements and (iii) susceptibility to abiotic and biotic stresses. The need (i) to take account of such variations in many disciplines of applied agrometeorology and in agricultural applications and (ii) for phytotronic studies to assess the physiological features of new cultivars is stressed.

In Chapter 3, methodology for delineation of global agroclimatic analogues for introduction of exotic crops and parasites/predators of crop pests and disease organisms is presented. Adoption and usefulness of the Dynamic Approach in assessment of life-duration, growth, developmental-rhythm and production and partitioning of drymatter of crops, the concept of Climatic Fertility and the need for and usefulness of crop-wise climatic zonation of irrigated crops are detailed.

In Chapter 4, the concepts of Solar Constant and Extraterrestrial Radiation, terminology used for expression of solar radiation intensity and their interrelationships are detailed. Daily values of Extra-Terrestrial Radiation on a weekly basis at 10 degree latitude intervals from 70 degrees South to 70 degrees North are presented. Dissipation of solar radiation during its passage through the atmosphere, energy exchanges at the Earth's surface and the resultant Net Radiation are discussed. Measurement, estimation and agricultural implications of net Radiation and its components are presented. Solar radiation relationships of crop yields and ways of coping with the anticipated reduction in solar radiation under global warming are explained.

In Chapter 5, the important role of water and the need for optimal and conjunctive use of water resources in tropical crop production are stressed. Field measurements for determining water consumption of crops are described. The process of evaporation from soils and crop transpiration in relation to agrometeorological factors are detailed. The temporal variations in evaporation and transpiration with growth of the crop and the concept of relative evapotranspiration and its use in segregating the influences of soil, crop and weather on crop water consumption are elaborated. Data requirements and methodology for (i) evaporimetric irrigation scheduling of clear season crops and (ii) assessment of supplementary irrigation needs of crops in rainy season are presented. Criterion for assessing water use-efficiency is given. Methods for savings in irrigation water, strategic use of groundwater and collection and re-use of surface runoff from rains are stressed.

In Chapter 6, after a review of methodology for delineation of Dryland Cropping Period (DCP), criteria in terms of minimum assured rainfall amounts at 50% probability level and Evaporative Power of Air (EPA), for a given distribution of rainfall are laid down. Demarcation and utility of homogenous zones based on DCP are presented. Collection and re-use of surface run-off from rains through climatologically designed on-farm-reservoirs is stressed. Climatological aspects of feasibility of rice cum fish culture are examined. A rainfall budgeting methodology for assessing the temporal march of soil moisture stress days for a given crop, soil type and rainfall distribution and determination of Crop Drought Intensity (CDI), thereof is presented. Data requirements for use of the methodology and procedures for preparation of CDI maps are outlined.

In Chapter 7, undesirable consequences of over-protection of crops against pests and diseases are pointed out. Besides direct effects of weather, the need to take cognizance of influences of weather-induced presence or absence of alternate hosts, carriers, vectors, parasites and predators and modifying role of crop factors like food supply, host resistance and growth stages in the incidence, multiplication and spread of pests and diseases is emphasised. The possibility of exotic origin of many pests and diseases and ways to avoid afflictions of pests and diseases are made out. Approaches for anticipation of pest and disease attacks, limitations and usefulness of laboratory and field studies in evaluating pre-disposing conditions for incidence, multiplication and spread of pests and diseases are detailed.

Thumb rules based on local studies for anticipation of a specific pest or disease at a given location, statistical analyses of concurrent data on pests/diseases and weather to derive crop cum weather based regression functions with lead-time for prediction/early warning and a dynamic approach in which weighted value is assigned to meteorological parameters depending on their level and importance are detailed. The concepts of Biofix, Economic Threshold Level of pest and Critical Disease Level (CDL), are presented and their utility in use of spore and insect trap data in initiation of timely control measures is highlighted. Methodology for use of routine temperature data in estimation of Leaf-Wetness Duration for real-time disease warning, difficulties in issue of mid-seasonal advisories for control operations against pests and diseases and crop-weather situations in which forewarnings of pests/diseases would be most effective are detailed. The need for weather cognizance in effective, chemical and biological control of pests and diseases is emphasised. Bright prospects for real-time forewarning of pests/diseases is made out.

In chapter 8, the physical processes involved in erosion of soil by rainfall and importance of kinetic energy of rainfall in it are brought out. Methodologies for computation of rainfall energy and rainfall erosivity thereof and indirect estimation of rainfall erosivity for preparation of iso-erodent maps of rainfall are presented. Process of removal of soil by wind, erodibility of soils and forms and types of soil erosion by wind are detailed. Influence of non-weather factors on wind erosion, computation of Wind Erosion Index and soil loss due to wind erosion are discussed. Measures for control of soil erosion by wind and determination of Predominant Wind Erosion Direction in organising control measures are presented.

Chapter 9 details observed/anticipated changes under global warming of (i) Indian Summer Monsoon and global rainfall (ii) concentrations of Carbon Dioxide, Methane and Nitrous Oxide in the atmosphere (iii) air temperatures (iv) cloudiness and solar radiation and (v) Evaporative power of air. The role of aerosols in mitigating global warming and controversies on climate change are examined. Effects of climate change on plant processes, crops, pests, diseases and weeds are succinctly dealt with. The need for assessment of crop prospects in future climate on a crop-wise, region-wise and season-wise basis in a holistic manner under pessimistic, optimistic and expected climatic scenarios is stressed. Combating the challenges of climate change on crops, mitigation of agricultural effects on climate change, ways of increasing sink capacity for CO_2 and carbon sequestration are discussed.

In chapter 10, the need for Agrometeorological Advisory Services (AAS), is highlighted. Requirements of providers and user-interests of AAS particularly a network of crop truth ground stations and infrastructure for speedy interactive communications are set out. Quick and unique processing of crop cum weather data for timely issue of agromet advisories and usefulness and effectiveness of AAS are detailed. Agrometeorological forecasting of unit area yields of rainfed and irrigated crops as part of AAS is dealt within an elaborate manner. Need for and modalities of operation of weather-based crop insurance schemes are spelt out.

Chapter 11 emphasises the reasons for and need to reverse current status of agricultural non-sustainability. Cooperative farming, optimal and conjunctive use of water resources, organic cropping, integrated management of pests and diseases, climate-based crop planning and fixation of Minimum Support Price for crop produce, weather-management of crops, breeding of cultivars with (i) physiological control to reduce water uptake under atmospheric and soil moisture stresses (ii) maximal interception of solar radiation and (iii) higher Radiation Use-Efficiency as corner stones for ensuring future agricultural sustainability with environmental conservation are detailed. Last but not the least, ways of ensuring food security to landless farm labourers through non-farm and off-season work programmes for creating permanent assets to aid future crop production are indicated.

It is hoped that the principles and methods enunciated in this book will get incorporated in the guidelines/manuals relating to agromet aspects of irrigation scheduling, crop yield forecasts, climate and rainfall zonations, assessment and depiction of crop droughts, soil erosion control, impact assessment and mitigation of effects of climate change.

- Author

Contents

CHAPTER 1
Origin of Developments Relating to and/or in Agricultural Meteorology

CHAPTER 2
Eco-Physical Aspects and Eco-Physiological Basis of Agricultural Meteorology

CHAPTER 3
Crop-Climate: Analogues and Zones

CHAPTER 4
The Concept and Relevance of Radiation Balance in Crop Culture

CHAPTER 5

Agrometeorology of Crop Water Usage

CHAPTER 6

Dry Farming Agrometeorology

CHAPTER 7

Avoidance, Anticipation and Control of Pests and Diseases

CHAPTER 8

Erosion of Top Soil by Rain and Wind

       8.1.1   Splash Erosion 199

       8.1.2   Rainfall Intensity 200

       8.1.3   Run-Off 201

       8.1.4   Scouring 201

8.2   Soil Erosion Models 202

       8.2.1   Prediction of Soil Loss 202

8.3   Rainfall Erosive Capacity 203

       8.3.1   Computation of Rainfall Energy 203

       8.3.2   Method of Computation of
                Rainfall Energy, R 203

       8.3.3   Need for basic Determination of
                Rainfall Energy 204

8.4   Erosion of Soil by Wind 205

       8.4.1   Erodibility of Soils by Wind 206

       8.4.2   Forms of Soil Erosion 206

       8.4.3   Types of Wind Erosion 207

       8.4.4   Influence of Non-Weather Factors 208

       8.4.5   Deposition of Soil Particles 210

       8.4.6   Estimation of Risk and Amount of
                Soil Loss by Wind Erosion 211

       8.4.7   Wind Erosion Equation 212

       8.4.8   Wind Erosion Measures 213

       8.4.9   Predominant Wind Erosion Direction 214

       References 215

CHAPTER 9

Climate Change

CHAPTER 10

Agrometeorological Advisory Services

CHAPTER 11

Agricultural Renewal and Sustainability

Abbreviations

AAS	:	Agromet Advisory Services
AFCY	:	Agrometeorological Forecasting of Crop Yields
ANN	:	Artificial Neural Network
DCWSM	:	Dynamic Crop-Weather Simulation Model
DCWBI	:	Dynamic Cumulative Weather-Based Index
FAO	:	Food and Agriculture Organisation
FASAL	:	Forecasting Agricultural Outputs Using Space, Agrometeorology and Land-Based Observations
FACE	:	Free Air Carbon Dioxide Enrichment
FATE	:	Free Air Temperature Enrichment
ISMR	:	Indian Summer Monsoon Rainfall
IMPD	:	Integrated Management of Pests and Diseases
IPCC	:	International Panel for Climate Change
MRWF	:	Medium Range Weather Forecasting
OFR	:	On-Farm-Reservoirs
OACUWR	:	Optimal and Conjunctive Use of Water Resources
P & D	:	Pests and Diseases
SPAR	:	Soil, Plant, Atmosphere Research Unit
SIN	:	Supplementary Irrigation Need
TSI	:	Transpiratory Satiation Index
UNEP	:	United Nations Environment Programme
WRSI	:	Water Requirements Satisfaction Index
WMO	:	World Meteorological Organisation

CHAPTER 1

Origin of Developments Relating to and/or in Agricultural Meteorology

In the 4th Century. Aristotle, whose "Meteorologica" elevated Meteorology from the realm of mythology to science, had observed "Annus Fructificat Non Terra" meaning it is the year (read weather) and not the soil that determined crop yields. Rural proverbs found in almost all Indian languages abound in giving thumb rules for (a) anticipation of local weather (b) timing of agricultural operations in light of expected weather and (c) advanced assessment of crop prospects on basis of realized weather. Basu (1953) found no scientific basis for the proverbial beliefs. Nevertheless, the rural proverbs reveal the anxiety of farmers to know in advance the likely weather situations for crop operations and yield assessment. Thus, the knowledge of farmers on influence of weather on crops is times immemorial.

Venkataraman and Krishnan (1992) in their book "Crops and Weather" have given an historic narration of the various topics of studies covered in the chapters of the book. It was felt that culling out from the above book and other sources, a chronological bird's-eye account of the origin of and developments in and related to agrometeorology can be made to inculcate in students of agrometeorology an appreciation of some pioneering work and concepts that have laid the foundation for many lines of work currently in use in applied agrometeorology.

1

In light of the above, some pioneering work on crop-weather studies, prior to acceptance in the 1930s of agrometeorology as a specialised science, are cited and their relevance to appropriate fields of applied agrometeorology currently in use are briefly mentioned. Initiation of important lines of work and significant developments are presented decade wise from 30s to first decade of the current century.

1.1   Pioneering Work

1.1.1   Degree-days

Reamur (1735) was the first to attempt a quantification of agrometeorological relationships by trying to relate duration of crop-life in terms of accumulated mean temperatures. Reamur's work remained dormant till it was resumed by Abbe (1905). The concept of Reamur of accumulated mean air temperatures have been extensively adapted in the 20th Century for understanding and/or explaining phasic development of plants. In these, accumulations (i) above a specified mean air temperature depending on crop (ii) night air temperature (iii) excluding values of temperature (a) above a specified maximum and (b) below a specified minimum (iv) product of photoperiod and temperature called Photothermal Units and (v) products of hours of bright sunshine hours and temperature called Heliothermal Units have been used.

1.1.2   Photoperiodism

The observations in the differences in flowering behaviour of avenue trees subject to the influence of street lights compared to those in the dark led Garner and Allard (1920) to the discovery of Photoperiodism. This concept led to the classification of plants as long-day, short-day and neutral types depending on their photoperiodic requirements for flowering. Garner and Allard 1920) themselves classified some plants which will not flower under photoperiods of less than 12 hours or more than 14 hours as intermediate types.

Flowering of crops on account of photoperiodism is agriculturally undesirable as it gave little scope for agronomic manipulation in management of crops. The need for removal of photo-sensitivity became crucial and was successfully done by plant breeders by evolving photo-insensitive varieties of many crops. Further findings, relating to photoperiodism and their agronomic and genetic applications, are given in Chapter 2. The main findings relating to photoperiodism are that

  (i) the continuity in night period was the operative factor

 (ii) the photoperiodic effect was exerted even at intensities as low as in moonlight

(iii) the red spectrum is the most active in photoperiodic induction

(iv) either mature leaves or growing tips of plants are the receptors of the photoperiodic stimulus and

 (v) the photoperiodic effect could be modified or even eliminated by influence of temperature, especially night temperatures.

1.1.3   Pests and Diseases

Potato famine in Europe and North America, due to Potato Blight, in the middle of 19[th]century and Chestnut Blight in USA by end of 19th century triggered the need for an understanding of weather relations of diseases. Studies of pests in relation to weather are complicated by the operation of factors like mobility and hibernation through which pests can avoid encountering adverse weather. The investigations in the 1920s pioneered the work on weather aspects of incidence of pests of rice (Mishra, 1920; Ghosh, 1921; Austin, 1923; Ramachandra Rao, 1925; Hegdekatti, 1927).

Pierce, et. al. (1912) identified some natural enemies of the Cotton Boll Weevil and demonstrated their usefulness in control of the Cotton Boll Weevil in U.S.A. They had, advocated adoption of a sound climatic approach for biological control of pests. It is worth nothing that neglect of the climatic aspect led to the failure of the attempt in biological control of the Sugarcane Mealy Bug in Egypt and Palestine (Uvarov, 1931). Biological control of pests and diseases through introduction, from exotic regions, of their natural parasites/predators based on the climatology of places and periods of their origin and introduction is now an accepted and widely used non-chemical method in the fight against crop pests and diseases.

1.1.4   Others

De Saussure can rightly be called the pioneer in the field of pathways in photosynthetic fixation of carbohydrates as he had hinted at the

Crassulacean Acid Meabolism pathway in photosynthesis as early as 1804 and the same was confirmed in 1892 by Aubert. The work of Money (1883) on the climatic needs of the Tea crop can be held as a pioneering study on the assessment of climatic requirements of crops. Edward Mawley, President of The Royal Met. Society had observed in 1890 that "there are few sciences so intimately connected with each other as Meteorology, Agriculture and Horticulture". The above set the tone and emphasised the need for development of Agricultural Meteorology as a scientific discipline.

Work of (a) Arnon (1909) and Wiley (1901) on the influence of temperature on flowering of Sugarcane and of latitudinal spread of Sugar beet respectively (b) Howard et.al. (1914) on the effect of environment on quality of Wheat (c) Bates (1911) on influence of windbreaks on crop yields and (d) Koppen (1918) on climatic classification, can be cited as pioneering studies in applied agrometeorology. The concept of Relative Transpiration of Livingston, (1906) can be considered the fore-runner in the meteorological determination of crop water needs. Soil temperature had been recognised as an ecological factor in early 1920s (Jones and Tisdale, 1921). The observation of Howard (1924) that wheat production in India is a gamble on temperatures gained as much attention as the adage that the Indian Budget was a gamble on the monsoon.

1.2   Recognition

International recognition of Agrometeorology as a scientific discipline came about in 1913 with the formation of the Commission for Agricultural Meteorology, CAgM, in 1913 by the then International Meteorological Organisation, IMO. Due to World War I, CAgM had to be reconstituted in 1918 and had its first meeting in 1923. H. Hooker, President of the Royal Meteorological Society had, in 1921, suggested the use of Statistics to correlate crop yields with weather variables. A crop weather scheme was begun to be operated in U.K. in the 1920s. As a result of the deliberations of The Royal Commission on Agriculture in India constituted in 1928, the Division of Agrometeorology came to be established in 1932 at Pune under the India Meteorological Department. This catalysed subsequent international developments in the field of Agrometeorology.

1.3 Developments

1.3.1 1930s

Studies on (a) profiles of temperature and humidity (i) over and close to many natural surfaces by German workers (Geiger, 1930) and (ii) inside crop canopies vis-a-vis that in the open by Indian workers (Ramdas et. al. 1934) (b) likely contribution of Dew to water needs of crops (Wadsworth, 1934) (c) use of January temperatures in anticipating incidence of Downey Mildew of Tobacco in Southern U.S.A, which may be deemed as a forerunner in forecasting diseases using the degree day concept (d) the assessment of the stage at which a pest caused economic damage (Pierce 1934), which later led to the development of the concepts of economic injury level and Biofix, were of a pioneering nature.

The need for a separate network of stations recording in a farm environment meteorological observations of interest to agriculture not recorded routinely at synoptic weather observatories, recording of observations at 0700 and 1400 hours Local Mean Time to obtain range of weather parameters from observations with only eye reading instruments were recognised early. Standardization of field lay out and sampling procedures, on a crop-wise basis, for (a) assessing initial and final population density (b) recording (i) phenological and phenometric crop attributes and (ii) yields and (c) determining extent and intensity of incidence of major pests and diseases formed the basis for operation of All India coordinated crop weather Scheme relating to many crops.

1.3.2 1940s

The concept of Vernalisation, developed in the 1940s can be considered a milestone in adaptive agrometeorology and a classic example of adoption of an agronomic strategy to overcome harsh weather conditions. In this seeds of winter wheat are exposed to a chilling temperature of 3 degrees centigrade for a definite period so as to induce shooting on sowing in the pre-winter period such that the seedlings will remain alive under a snow cover in winter, resume growth on advent of spring to complete their life cycle.

Work of (i) Burton, (1941), on the influence of weather factors on viability of seeds in situ and in storage, (ii) Penman (1941) and Staple and Lehane (1944) on weather aspects of evaporation from bare soil

(iii) Mehta (1940) on the exotic origin of wheat Rusts and (iv) Parija (1943) on high temperature vernalisation by pre-sowing treatment of seeds to modify phasic development of plants were notable contributions. The laboratory work of Pradhan (1946) on growth and development of pests leading to the concept of a Biometer in which the temperature graduations in a thermograph are replaced by lines of average development of the pest for a given stage at that temperature and the enunciation, in 1948, of the concept of Potential Evapotranspiration, PET, in USA (Thornthwaite, 1948) and in U.K. (Penman, 1948) and the construction of the first Phytotron at the Earhart Laboratory of the California Institute of Technology in 1948 were landmark developments in the 1940s. The concept of PET has later helped (i) to unify apparently diverse data on crop water consumption and needs (ii) in formulation of weather-based irrigation scheduling of crops and (iii) in isolating and quantifying the role of soil and crop factors on crop water requirements. The phytotron helped in carrying out of intensive studies on the influence of weather factors on growth and development of plants. Enunciation of the concept of Ecological Geography and identification of global agroclimatic analogues (Nuttonson, 1949) for many regions of the world for diverse crops was a significant development. The feasibility of forecasting of Potato Blight from weather data (Cook, 1949) was the fore-runner for weather-based anticipation of and assistance in control operations against many crop diseases.

1.3.3   1950s

The coming into being of the World Meteorological Organisation, WMO as a specialised agency of United Nations in 1951 and inclusion of the Commission of Agricultural Meteorology CAgM as one of the eight technical commissions of WMO were significant developments that fostered global development of scientific and service aspects of Agricultural Meteorology.

Discovery of the Calvin cycle in formation of photosynthates leading to the identification of crops as C3 types was an important development of the 1950s. From studies in climate-controlled chambers, emerged another important discovery, namely the concept of Thermoperiodicity (Went, 1957) i.e., the differential response of crop species to daytime, night-time and mean air temperatures (examples: Solanaceae to night temperatures; Papillionaceae to daytime temperatures and Graminaceae to mean air temperatures). Thermoperiodicity should rank in importance equal to the discovery of photoperiodism in crops in the 1920s. Other significant development in the 1950s, were: (a) use of monomolecular films of long-chain alcohols for reducing evaporation from free water surfaces (Mansfield, 1958) (b) development of various types of lysimeters to measure daily evapotranspiration of aerobic and anaerobic crops (c) opening up of the feasibility of forecasting of pests and diseases from weather data(Bourke, 1955) and (d) enunciation of the concept of Economic Threshold Limit, ETL for initiation of control action against pests using data from insect traps.

The beginning of Dynamic Crop Weather Simulation modeling can be traced to the 1950s due to publication of   pioneering papers on (i) plant development (Monsi and Saeki,1953) and (ii) plant modelling (Dewit, 1959) and (iii) fundamental work on plant photosynthesis and crop productivity (Davidson and Philip. 1958; Went, 1958).

1.3.4   1960s

Development of the first model on water transfer in the soil-plant-atmosphere continuum by Gardner (1960) and development of models for (i) assessment of photosynthesis of crop canopies (De Wit, 1965), which was the fore-runner in the development of Dynamic Crop-Weather Simulation Models (ii) concepts initiated in the 1950s (Ross, 1966) leading to the expression in mathematical terms of (a) plant growth and (b) plant responses to environmental conditions (Duncan et. al. 1967) need special mention.

The launching of the Polar Orbiting satellites gave a boost to development of Satellite Agricultural Meteorology as one could observe any place on Earth and view every location twice each day with the same general lighting conditions due to the near-constant local solar time. Being closer to earth the Polar Orbiting Satellites gave scope for a much better resolution than their geostationary counterparts.

The Green Revolution brought about by the breeding and use of short-statured hybrid varieties of many grain crops leading to quantum jumps in unit area yields of many crops was the most important development of the 1960s.

The world's first truly operational Geographical Information System, GIS was developed in Canada in 1960. GIS later came to be used extensively to receive satellite-sensed agrometeorological data and produce outputs of agrometeorological importance. Formation in 1964 of the Laboratory for Computer Graphics and Spatial analysis and development of a number of important theoretical concepts in spatial data handling were important developments.

Since pan evaporation integrates the effects of all factors affecting water needs of crops, its role in crop water management is obvious. However, evaporimeters in various countries vary in dimensions, material of construction and manner of maintenance, mounting and measurement. Their readings were vitiated by bird visitations. Thus, the programme for international comparison of evaporimeters drawn up by the Commission of Instruments and Methods of Observation, CIMO of WMO in 1964 involving principally the USA Class A Pan, Russian GGI 3000 evaporimeter and the Russian 20 m^2 tank was an important endeavor. Evaporation from the Russian 20 m^2 tank represented evaporation from a shallow lake which is equivalent to Potential Evapotranspiration.

An Interagency Group on Agricultural Biometeorology was formed by FAO, WMO and UNESCO. Rigorous field studies on physical processes in crop environment like fluxes of momentum, energy and moisture and intensive phytotronic studies on physiological processes of crops had been undertaken. The conservative nature of yield per day per unit area of crop cultivars in a given season was established (Swaminthan, 1968). The mathematical breakthrough of Incomplete Gamma Distribution (Thom, 1966) helped in analyses of rainfall amounts in short time-units and in determination over short time-units (i) minimum assured rainfall at a given probability level and (ii) probability of realising a given amount of rainfall and was a significant contribution in the field of dryfarming meteorology. The discovery of the Hatch-Slack photosynthetic pathway led to identification of plants as C4 types. Enunciation of the concept of Critical Disease Level, CDL which relates to the stage of development of the disease before which control operations are not required and after which control operations will be ineffective (Shoemaker and Lorbeer 1969) helped in timely and effective control operations against crop diseases based on data of spore traps and was another major development.

1.3.5   1970s

Appointment of a Rapporteur by CAgM of WMO in 1974 to study the application of remote sensing techniques to solving of agrometeorological problems was an important step that gave a boost to the development of satellite-sensed agromet parameters and GIS applications in agrometeorology. Analyses of data of about 10 years, recorded under the International Comparison of Evaporimeters, brought out the fact that the ratio of evaporation from pans to that from the Russian 20 m² tank varied with the rate of evaporation. This revealed, that for use of Pan Evaporimetry for scheduling irrigation of crops, ratio of EP to PET must be determined on a season-wise and region-wise basis. Standardisation of procedure by FAO for computation of Wind Erosivity Index was an important step.

Prior to 1970 crop-weather relationship studies used to be conducted in Phytotrons or environmental chambers in which the ambient conditions were kept constant and hence quite different from the macro and micro-climatic conditions experienced by crops in the field. In field crop trials, phenological and phenometric attributes could be related to positive and negative deviations from normal values of all meteorological parameters except Carbon Dioxide in which higher values never prevailed in the past. In view of the expected increase in Carbon dioxide concentration under global warming, it was necessary to study gas exchanges between the crop and its environment under higher level of Carbon Dioxide. In this context, a beginning was made by Heagle et. al. (1973) to study plant processes under field conditions through the use of Open Top Chambers, (OTC). The OTCs are plastic enclosures, with an open top, constructed of an aluminum frame covered by panels of PVC plastic film. Air enriched with CO_2 is pulled into the bottom of the chamber, and then blown through the open top of the chamber. OTCs, inexpensive and easy to construct and maintain, are not suited for the study of large vegetation.

The next development in the study of gas exchanges in natural cropped conditions was the design and construction in 1978 of the Soil-Plant-Atmosphere Research System SPAR, by Phene et. al. (1978) at Florence, South Carolina, USA. In SPAR, the area earmarked for the study is allowed to receive natural sunlight. Soil and aerial environments are precisely controlled in the earmarked area and gas exchanges are rapidly and automatically measured under natural sunlight with attendant variations in solar radiation intensity and its spectral distribution. The latter features of solar radiations are difficult to be obtained in artificially lit Phytotrons and growth-chambers.

International crop research institutes with agerometeorological units were established under the Governing Council for International Agricultural Research, GCIAR, to meet needs of specific climatic zones and to analyze crop-weather data gathered on specific crops on a global scale. There was a spurt in the activities of FAO in Agrometeorology under its Environment Natural Resources and Services Division. The FAO scheme on agrometeorological monitoring and forecasting yields of rainfed crops based on rainfall budgeting and water satiation concept was started. The separation of the evaporative component of evapotranspiration under incomplete crop cover and irregular wetting of soil surfaces (Ritchie, 1972) was a land mark development. The Satellite Instructional and Television Experiment, SITE, in India in which agricultural, agrometeorological and meteorological experts jointly reviewed the weekly field-position of crops along with concurrent meteorological data and the anticipated weather, to issue advisories for crop operations, became the fore-runner for initiation of Agromet Advisory Services in India. Methodology for assessment of wind erosivity index was standardised by FAO. A computerised system for forecasting crop diseases was initiated (Castor et. al. 1975). Assessment of areal spread and health of crops by satellite imagery was commenced. Studies on Radiation Use-efficiency and influence of elevated CO_2 and shading on crop yields, which have a bearing on assessment of crop prospects in a warm climate, were initiated.

The development and use of the concept of Biofix (Riedl et al. 1976), which relates to the date of occurrence of a particular event in the life of an organism was an important development leading to the use of catches in insect and spore traps for minimal and effective control operations against pests and diseases.

Continuation of the work on Dynamic crop weather simulation model (Dewit et. al. 1970; Curry 1971; Dewit and Goudrian, 1974) led to the formulation of Dynamic Crop weather Simulation Model, on Sorghium by Arkin et. al. (1976).

1.3.6 1980s

Formation of The Intergovernmental Panel on Climate Change (IPCC), an international organization with a mandate "to assess on a comprehensive, objective, open and transparent basis the best available scientific, technical and socio-economic information on climate change" initially by WMO and UNEP in 1988 and later endorsed by United Nations , was an important development.

WMO and FAO actively collaborated on (a) provision of meteorological aids for prevention of desertification (b) application of climatic data for effective irrigation planning and management and (c) weather-based forecasting of crop yields. WMO standardised the procedure for computation of rainfall erosivity. Flooded rice was identified as a significant source of atmospheric Methane. Noticing of the increase in difference in yield in farmers' fields with that of a nearby research station with increase in rainfall (Sivakumar et. al. 1983) highlighted a serious deficiency in the lab to land transfer of technology. Delineation of the role of crop physiology in water uptake by crops, studies on influence of CO_2, solar radiation etc., having a bearing on impact of climate change on crops, suggestion for use of decreased respiration rates as a selection criterion in breeding crops for higher yields (Wilson and Jones, 1982) were significant features. The later part of 1980s witnessed a spurt in development of crop models (Sinclair, 1986; Hammer et. al. 1987; William et. al. 1989). Dynamic Crop Weather Simulation Models, DCWSMs were developed and field tested for many important crops. Satellite data were also used to improve output of crop yields from DGWSMs.

An important development was the plethora of Free Air Carbon Dioxide Enrichment (FACE) experiments pioneered by Rogers et al. (1983) and Jones et. al. (1984), who adopted the fumigation methodology (Greenwood, et. al. 1982) used to study effects of air pollutants on crops and adapted the SPAR Technology (Phene et al. 1978). In the FACE experiments, gassing circles of sufficiently large diameter are installed in a field. Inside the circles, the aerial CO_2 concentration can be increased with reference to the natural ambient CO_2 concentration, with automatic control of wind speed and wind direction to maintain the experimental concentration level. In such a set up, Carbon Dioxide is blown from a tank through vertical standing vent pipes into the exposure area. Sensors measure wind speed, wind direction, and CO_2 concentration and a computer control system regulates and monitors the CO_2 releases. The above felicitated the studies on effects of rising levels of CO_2 on crop yields.

The first genetically modified plant, an antibiotic-resistant Tobacco cultivar, was developed in 1983. However, the activity of genetic modification of plants remained dormant for another ten years.

1.3.7 1990s

There was a sudden spurt in evolving of genetically modified, GM plants. GM cultivars for a number of crops were developed in this decade. A very important development in the 1990s was the successful use of high resolution satellite imagery for crop identification, crop area determination, estimation of moisture content of top layers of soil, determination of crop emergence and hence of sowing dates, yield-determining crop attributes, crop phenology, crop moisture deficit etc. These enabled the preparation of crop inventories at periodic intervals on a crop-wise and region-wise basis and led to formulation of methodologies for agrometeorological forecasting of gross, regional crop yields through combination and integration of ground-truth and satellite sensed agrometeorological data (Gommes 1998;Moulin et al. 1998).

The other notable developments in the decade were (a) initiation of in-depth studies on an extensive global scale on (i) the likely changes in direction and magnitude of changes in weather parameters on account of the Greenhouse Effect of anthropogenic origin (b) effects of climate change on crop production in future (c) ways for mitigation of the harmful effects of (i) climate change on crops and (ii) agricultural practices on global warming (d) standardisation of methodology by FAO for determination of temporal march of Evapotranspiration of given crops at given locations and irrigation scheduling thereof (e) decreasing trends in observed Pan Evaporation and computed Reference Crop Evapotranspiration despite increase in temperatures in many parts of the world and (f) studies on likely reductions in water needs of crops in future climate.

1.3.8 First Decade of the Current Century

The decade has been one of achievements and concern as detailed below.

Much of the work mentioned above were being carried in an intensive manner and extensive scale through adoption/adaption of pioneering developments. In this, mention needs to be made of (a) improvements in the provision and end-use of crop-weather agrometeorological advisory services (b) forecasting of regional yields of many irrigated and rainfed crops by use of satellite data/imageries of an agrometeorological nature (c) studies relating to (i) assessment of the adverse influences of climate change on yields of many crops in many regions and seasons (ii) reduction of emission of Methane from flooded rice and GHGs from agricultural operations and (iii) increased Carbon sequestration.

A decreasing trend in rate of increase in crop production of even irrigated crops in the last 15 years due to pollution of air, degradation of soil and contamination of water resources on account of over-use of water and inorganic agrochemicals, production of all crops much in excess of national requirements and high market prices of all agro-industrial crop produce have brought to the fore the urgent need for climatic cognizance in production and protection of crops to ensure agricultural sustainability with environmental conservation and planned, specialised and regionalised production of crops in an economic manner. In this, inadequacy of apparent solutions like GM cropping and national rivers' linking have been brought out. Cooperative farming, organic cropping, optimal and conjunctive use of water resources and integrated management of biotic stresses have emerged as the solutions.

Progress of international collaboration in reduction of green house gases as per committed time bound targets in various regions has been stymied due to the concept of trading in greenhouse gas stocks and dilemma of developing countries' attempts in ensuring development with reduced GHG emissions.

References

Abbe, C. 1905. A first report on the relation between climate and crops. U.S. Dept. Agri. and Weather Bureau, Bulletin No. 36. 386pp.

Arkin, G.F.; Vanderlip, R.L. and Ritchie, J.T. 1976. A dynamic grain sorghum growth model. Trans. Amer. Soc. of Agricl. Engineers. 19: 622-630.

Arnon, I. 1909. (Cited by Coleman, 1963).

Austin, G.D. 1923. The paddy fly (Leptocorisa varicornis). Crop Agric. 60: 118-119.

Basu, S. 1953. Weather lore in India. Ind. Jl. Meteorol. and Geophysics, 4: 3-12.

Bates, C.G. 1911. Wind breaks : Their influence and value. U. S. Deptt. Agric. Forest. Serv. Bull. 86: 100pp.

Bourke, P.M.A. 1955. The forecasting from weather data of potato blight and other plant diseases and pests. Tech. Note 10, World Met. Org. Geneva. 48pp.

Burton, L.V. 1941. Relation of certain air temperatures and humidities to viabilities of seeds. Contrib. Boyce Thompson Instt. 12: 85-102.

Castor, L.L.; Ayers, J.E.; Menabb, R.A. and Krause, R.A. 1975. Computerized forecast system for Stewart's bacterial disease on corn. Plant Dis. Rep. 59: 533-536.

Coleman, R.E. 1963. Effect of temperature on flowering in sugarcane. Int. Sug. Jl. Res. 65: 351-353.

Cook, H.T. 1949. Forecasting late blight epiphytotics of potatoes and tomatoes. Jl. Agric. Res. 78: 545-563.

Curry R.B. (1971). Dynamic simulation of plant growth. Development of model. Trans. ASAE, 14: 946–949.

Davidson, J.L. and J.R. Philip, 1959. Light and pasture growth, 181-187 pp. In Climatology   and   microclimatology. Proc. Canberra Symp. 1956, Unesco, Paris.

De Wit, C.T. 1959. Potential photosynthesis of crop surfaces. Neth. Jl. Agric. Sci., 7: 141–149.

De Wit, C.T. 1965. Photosynthesis of leaf canopies. Agricl. Res. Rept. 663: 1-57 pp. Pudoc, Wageningen, The Netherlands.

De Wit, C.T. and Goudriaan J. 1974.   Simulation of Ecological Processes. Wageningen, Pudoc. 159 pp.

De Wit, C.T.; Brouwer R. and Penning de Vries F.W.T. 1970. The simulation of photosynthetic systems In: Setlik I. (Ed). Prediction and measurement of photosynthetic productivity. Wageningen. The Netherlands: Centre for Agricultural Publishing and Documentation, 47–70.

Duncan, W.G.; Loomis, R.S.; Williams, W.A. and Hanau, R. (1967). A model for simulating photosynthesis in plant communities. Hilgardia, 38: 181–205.

Gardner, WR. 1960. Dynamic aspect of water availability to plants. Soil Science, 89: 63-73.

Garner, W.W. and Allard, H.A. 1920. Effects of relative lengths of day and night and other factors of the environment on growth and reproduction in plants. Jl. Agric. Res. 8: 553-606.

Geiger, R. 1930. Mikroklima und Pflanzenklima. Handbuch der Klimatologie. Band 1. Teil D. Gebruder Borntraeger, Berlin. 46 pp.

Ghosh, C.C. 1921. Supplementary observations on borers in sugarcane, rice etc. Proc. 4[Th] Entomology meeting, Pusa. Govt. of Bengal, 105 pp.

Gommes, R. 1998. Agrometeorological crop yield forecasting methods. Proc. International Conf. On Agricultural Statistics, Washington, 18-20 March. Eds. Theresa Holland and Marcel P.R. Van Den Broecke. International. Statistical Institute. Voorburg, The Netherlands. 133-141 pp.

Greenwood, P. et al. 1982. A computer-controlled system for exposing field crops to gaseous air pollutants. Atmospheric Environment, 16: 2261-2266.

Hammer, G.L.; Woodruff, D.R. and Robinson, J.B. 1987. Effects of climatic variability and possible climatic-change on reliability of wheat cropping: a modeling approach. Agricultural and Forest Meteorol. 41: 123-142.

Heagle, A.S.; Body, D.E. and Heck, W.W., 1973. An open top field chamber to assess the impact of air pollution on plants. *Environ. Quai.*, 2: 365-368.

Hegdekatti, R.M. 1927. The rice gall midge in North Kanara. Agric. Jl. India 22: 461-463.

Howard, A. 1924. Crop production in India- A critical study of its problems. Oxford Univ. Press London. 200 pp.

Howard, A.; Leake, H.M. and Howard, G.L.C. 1914. Memoirs of the Dept. of Agriculture. Indian Bot. Ser. 5: 49.

Jones, F.R. and Tisdale, W.B. 1921. Effect of soil temperature upon the development of nodules on the roots of certain legumes. Jl. Agric. Res. 22: 17-32.

Jones, P.; J. Allen, L H.; K.W. Jones.; K.J. Boote and W.J. Campbell. 1984. Soybean canopy growth, photosynthesis and transpiration responses to whole-season carbon dioxide enrichment, Agronomy Jl. 76: 633-637.

Koppen, W. 1918. Klassification der Klimate nach temperature, Niederschlag and Jahresveriang. Petromannas   Mitt.   64: 193-203 and 243-248.

Livingston, B.E. 1906. The relation of desert plants to soil moisture and to evaporation Carneige Instt. Publication No. 50, Washington, D.C. 78 pp.

Mansfield, W.S. 1958. Reduction of evaporation of stored water. Proc. Symp. Climatology and Microclimatology. Arid Zones Research XI. UNESCO, 61-64.

Mehta, K.C. 1940. Further studies on the control of rusts in India. Indian Council of Agricultural Research, New Delhi, Monograph, No. 14, 224 pp.

Mishra, C.S. 1920. The rice leafhoppers (Nephotettix bipunctatus Fabr. and Nephotettix apicalis Motsch.). Memoir Dept. Agric. India. Entomology Series 5(5): 207-239.

Money, E. 1883. The cultivation and manufacture of tea. 4th Edition, Whittingham and co. London 184-193.

Monsi M. and Saeki T. (1953). Uber den Lichtfaktor in den Pflanzengesellschaften. Jpn. Jl. Botany, 14, 22–52. (In German).

Moulin, S.; Bondeau, A. and Delecolle. R. 1998. Combining agricultural crop models and satellite observations: from field to regional scales. Internatl. Jl. Remote Sensing, 19: 1021-1036.

Nuttonson, M.Y. 1949. Ecological crop geography of Germany and its agroclimatic analogues in North America. AICE, Washington, USA, 28 pp.

Parija, P. 1943. On the pre-sowing treatment and phasic development. Curr. Sci. 12: 88-89.

Penman, H.L. 1941. Laboratory experiments on evaporation from fallow soil. Jl. Agric. Sci. 31: 454-465.

Penman, H.L. 1948. Natural evaporation from open water, bare soil and grass. Proc. Roy. Soc. London (Ser. A), 193: 120-145.

Phene, C.J. et. al., 1978. SPAR-A Soi-Plant-Atmosphere Research System. Trans. Amer. Soc. Agricl. Engineers. 21: 925-930.

Pierce, W.D. 1934. At what point does insect attack becomes damage? Entomological News. 45: 1-4.

Pierce, W.D.; Cushman, R.A. and Hood, C.E. 1912. The insect enemies of the cotton boll weevil. U.S. Dept. Agric. Bur. Entomology Bull. 100: 1-99.

Pradhan, S. 1946. Idea of a biograph and biometer. Proc. Nat. Inst. Sci. India. 12: 301-314.

Ramachandra Rao, 1925. The silver shoot disease of paddy. Madras Agri. Dept. Year Book: 6-8.

Ramdas, L.A.; Kalamkar, R.J. and Gadre, K.M. 1934. Agricultural Meteorology. Studies in microclimatology. I. India Jl. Agric. Sct. 4: 351-467.

Reamur, R.A.F. de 1735. Observation du thermometer faites a Paris pendant l'annee 1735, compares avec cells qui ont ete faites sous la ligne a l'isle de France a larger et en quelques unes de nos Isles de l'Amerique. Paris Memoirs Academy Sci. 545 pp.

Riedl, H.; B. A. Croft., and Howitt, A.J. 1976. Forecasting codling moth phenology based on pheromone trap catches and physiological-time models. Can. Entomol. 108: 449-460.

Ritchie, J.T. 1972. Model for predicting evaporation from a row Crop with incomplete crop cover. Water Resources Res. 8: 1204-1213.

Rogers, H. H.; Heck, W. W. and Heagle, A. S. 1983. A Field Technique for the Study of Plant Responses to Elevated Carbon Dioxide Concentration. *Journal of the Air Pollution Control Association* 33: 42-44.

Ross J. (1966). About the mathematical description of plant growth. DAN SSSR 171 (2b), 481–483 (in Russian).

Shoemaker and Lorbeer, J.W 1969. Timing protection spray initiation to control onion leaf blight Phytopathology, 67: 402-409.

Sinclair T.R. 1986. Water and nitrogen limitations in soybean grain production. 1. Model development. Field Crops Research, 15: 125-141.

Sivakumar, M.V.K.; Singh, P.l. and Williams, J.S. 1983. In: Alfisols in the Semi-Arid Tropics: A Consultant's Workshop, 1-3, December, ICERSAT centre, India. 15-30 pp.

Staple, W.J. and Lehane, J.J. 1944. Estimation of soil moisture from meteorological data. Sci. Agric. 32: 36-47.

Swaminathan. M.S. 1968. Genetic manipulation of productivity per day. Special Lecture, Symposium on Cropping Patterns in India. Indian Council of Agricultural Research.

Thom, H.C.S. 1966. Some methods of climatological analysis. Tech. Note, 81, WMO. 53 pp.

Thornthwaite, C.W. 1948. An approach towards a rational classification of climate. Geogr. Rev. 37: 87-100.

Uvarov, B.P. 1931. Insects and climate. Transac. Entomol. Soc. London. 79: 1-247.

Venkataraman, S. and Krishnan, A. 1992. Crops and Weather. Publication Indian Council of Agricultural Research, 586 pp.

Wadsworth, H.A. 1934. Light showers and dew as a source of moisture for the cane plant. Hawaiian Planters Record, 38: 257-264.

Went, F.W. 1957. Experimental control of plant growth. Chronica Botanica 17. Ronald Press Co. New York. 343 pp.

Went F. W. (1958). The physiology of photosynthesis in higher plants. Preslia, 30: 225– 249.

Wiley, H.W. 1901. The sugar beet: Culture, seed development, manufacture and statistics. FMR. Bull. 52. U.S. Dept. Agric. Washington, 48 pp.

Williams J. R.; Jones C. A.; Kiniry J. R. and Spanel, D. A. 1989. The epic crop growth-model. Trans. ASAE 1989, 32: 497-511.

Wilson and Jones, J.G. 1982. Effect of selection of dark respiration rate of mature leaves on crop yields of Lolium-Perenne cv S23. Annals of Botany. 49: 313-320.

CHAPTER 2

Eco-Physical Aspects and Eco-Physiological Basis of Agricultural Meteorology

It has often been noticed that findings based on experiments with individual plants in phytotrons or in open areas in pots are not directly applicable for crop-field use. The above limitation arises from the fact that when plants get established as a community, either as crops or as natural vegetation, they exert influences quite distinct from that of an individual. For example, field populated to maximum density such that each individual plant yields the maximum appropriate to the prevailing weather conditions, will not give bumper yields. On the other hand the yield per plant in a field giving maximum yields under the same conditions will be less than optimal. A denser field will absorb a greater fraction of solar radiation and atmospheric precipitation. Again, crop water needs of a sparse field will be higher than a denser one. This leads to the paradox of increasing competition from plants leading to higher unit area yields and better use efficiencies of water and solar radiation.

Crops are known to exhibit marked similarities despite variations in the climate of their habitat. For example, the highest unit area yields of sugarcane are obtained in the Hawaiian Islands and the rainless coastal belts of Peru and cane growing periods of 24 months obtain in Hawaii, Peru and South Africa (Venkataraman and Krishnan, 1992). The above areas are climatically quite different and distinct from each other. Again, different crops exposed to the same weather vagary i.e., deviation from the normal, react differently with respect to (i) physical damage (ii) deviations from normal growth, development and yield and (iii) susceptibility to pest and disease afflictions. The above features can be traced to influences exerted by crop morphology, crop growth stages and by far the most important crop physiology.

2.1 Micro-Climates

When crops grow they begin to interfere with radiation, moisture and momentum fluxes between the earth and its overlying air layers and soon make the crop canopy as the active surface in these exchanges. Due to the above interference and transpiration from the leaf-profiles, its micro-climate i.e., the direction and gradients of wind-speed, temperature, humidity, radiation and illumination inside a crop are quite different from the open micro-climate of adjacent bare soils exposed to the same weather.

2.1.1 Dependent and Topo Micro-Climates

The in-situ micro-climates of crops are affected by phenomenon originating from or associated with natural surfaces such as sea breezes, katabatic winds, winds blowing through gaps in mountains etc., Geiger (1965) refers to such type of features as "Dependent Micro-Climate". Crop micro-climates are also influenced by topographic variations such as slope-gradients, orientation of slopes and nature of physical separation of the sloping surfaces. The influence of terrain on micro climates constitutes the realm of "Topo Micro-Climatology". Variations in photoperiod with direction and steepness of slopes, orographic influences of rainfall, funneling of cold winds into valleys are some examples of terrain effects on micro-climates. Dependent and topo micro-climates strongly influence the distribution of natural vegetation. Thus, vegetation in high altitude regions of tropics and temperate areas are comparable to those of temperate and polar areas respectively. Study of weather relations of crops growing in undulating terrain is a much more complex affair than those relating to plain areas.

2.1.2 Micro-Climates of Natural Vegetation

In situ, dependent and topo micro-climates of natural vegetation types determine the kind of flora that can grow in association with them. Because of this, specific natural vegetation zones are the habitat for medicinal plants and the original genetic stock of many crops and their cultivars currently in use. The need for conservation and careful preservation of our valuable heritage of biodiversity flows from the above.

2.1.3   Crop Micro-Climates

The micro-climate of crops has a great bearing on incidence of pests and diseases. Thus, by denying the ideal micro-climate to pests and diseases development though mixed cropping, many a times, incidence of pests and diseases could be avoided. Many plantation and spice crops require micro-climate provided by trees for optimal growth and yield. The influence of micro-climate is reflected in the compatibility and productivity of crops in mixed stands.

2.2   Influences of Crop Morphology

2.2.1   Drymatter Production

Several studies have shown that the total drymatter produced is directly proportional to the amount of Photosynthetically Active Radiation, PAR intercepted by it. Now, PAR is 0.45 times Global Solar Radiation, GSR (Monteith, 1965). The number of plants required per unit area of land to ensure optimal development of leaf surfaces in the quickest possible time and maximal absorption of solar radiation in a given environment depends on (i) the leaf architecture of individual plants, namely of leaf layers, number, dimensions and angle of orientation of leaves in each leaf- layer and (ii) capacity for adventitious growth like tillering and branching.

2.2.2   Crop Water Relations

The foraging capacity of roots for soil moisture varies with soil types for the same crop and in the same soil varies with crop types. The differences in root foraging capacities of plants lead to differences amongst (i) dryland crops in susceptibility to droughts under same rainfall regime and (ii) irrigated crops in quantum and interval of irrigation. Most of the trees have the stomata on the underside of leaves and this helps in reduced water requirements of tree crops and their survival in harsher soil moisture regimes.

2.2.3   Atmospheric Methane

Continued standing water on the puddled rice field creates hypoxic soil conditions under which rice cannot thrive. Rice roots require oxygen. Under flooded conditions 30 to 40% of the cortex around the central stele disintegrates to form aerenchyma cells (air pockets) that facilitate oxygen to diffuse to the roots. Thus, energy which should have been utilized for grain production is spent in development of above air pockets. Puddling leads to anaerobic decomposition of organic matter and production of Methane, a key constituent of Green House Gases (GHGs). However, the aerenchyma cells of the rice plant provide the conduit for 70% of the Methane released from rice puddled rice fields.

2.3   Influences of Crop Growth Stages

2.3.1   Susceptibility to Abiotic Stresses

Abiotic stresses refer to the negative impact of non-living factors on growth, development and yield of crops. Factors like soil moisture inadequacy, high and low temperatures, high winds and heavy rains constitute abiotic stresses. Susceptible crop stages to soil moisture stress varies amongst crops (Doorenbos and Kassam, 1979). The concept of Critical Sunlight Period (Ronald and Cicerone, 2002) stipulates that reduction in bright hours of sunshine during specified crop stages will reduce crop yields. The critical stages vary amongst crops. For Rice it is the 6 week period following Panicle initiation during which reduction in bright sunshine hours leads to reduced grain number (Swain et al. 2007). As a considerable fraction of drymatter grain comes from post-floral photosynthesis (Watson. 1956) for many grain crops, radiation and temperature during the flowering period of grain crops become important.

2.3.2   Susceptibility to Biotic Stresses

Biotic stresses refer to the harmful impact of living entities like pests, diseases and weeds. Susceptible growth stages to a given biotic stress vary amongst crops and for the same crop are different for different biotic stresses. In case of pests and diseases, they become injurious to crops at particular stage(s) of their growth. Thus, juxtapositioning of infective stage of a pest/disease and susceptible crop stage is necessary for infection to occur.

2.3.3   Reception of Stimuli

Photoperiod, in addition to its effect on flowering, may (a) exert other effects (Venkataraman and Krishnan, 1992). like (b) inculcation and breaking of bud dormancy and winter-hardiness and (c) influence (i) elongation and branching of stem (ii) shape, size, succulence and abscission of leaves (iii) formation of tubers, tuberous roots and bulbs (iv) fruit-setting (v) sex expression and (vi) germination The part of the plant that is receptive to photoperiodic induction varies amongst crops. In most of the plants, mature leaves are the receptors while in a few the stimulus is exerted through the growing point. Too high or too low a temperature in the photo-induction period may partly or wholly off set the influence of photoperiod. Thus, juxtapositioning of the receptive crop growth stage and appropriate photo and thermal regimes are required for photoperiodic effect to manifest itself.

To a lesser extent, temperature can also give rise to morphogenic effects, Unfavourable temperatures leading to slower development of leaf surface has permanent deleterious effect when occurring in the young seedling stage but not later.

2.4   Influences of Crop Physiology

As indicated at the outset, right from germination to harvest, crops exposed to the same environment react differently on account of differences in their physiological response to environmental factors. These are detailed below.

2.4.1   Germination

Most species of seeds are facilitated in their germination by the alternating temperatures that obtain in nature. Seeds of flower species do well under constant temperatures (Richards et al, 1952) but natural variations centered around the mean optimal temperature do not affect their germination. Some seeds of plants like Tobacco have a light requirement for germination and depth of placement in those cases affects germination. The above sometimes call for the pre-treatment of seeds.

2.4.2   Phasic Development

2.4.2.1 The Temperature Factor

Mention of the existence of Thermoperiodicity i.e., intra-special differences in plant responses to daytime, night-time and mean temperatures and accumulated temperatures as a measure of development has been made earlier. The base values above which mean, daytime or night-time temperatures have to be cumulated varies amongst crops range from 14 °C for Muskmelon to 2.2 °C for Spinach. For agricultural crops the range is from 4.5 °C degree for Wheat to 10 °C for Maize. The reductions in field-life-duration of crops for a degree centigrade for crops of differing base temperatures from 10 to 4.5 °C in different temperature regimes are shown in Table 2.1.

Table 2.1 Reduction in crop life-duration due to rise in temperatures

Mean Temp. Deg. C	Base Mean Air Temperatures																	
	10 °C			9 °C			8 °C			7 °C			6 °C			4.5 °C		
	Increase in Mean Air Temperatures-Degrees Centigrade																	
	+1	+2	+3	+1	+2	+3	+1	+2	+3	+1	+2	+3	+1	+2	+3	+1	+2	+3
15	17	19	37	14	25	33	13	22	30	11	20	27	10	18	25	9	16	22
16	14	25	33	13	22	30	11	20	27	10	18	25	9	17	23	8	15	21
17	13	22	30	11	20	27	10	18	25	9	17	23	8	15	21	7	14	19
18	11	20	27	10	18	25	9	17	23	8	15	21	8	14	20	7	13	18
19	10	18	25	9	17	23	8	15	21	8	14	20	7	13	19	7	12	17
20	9	17	253	8	15	21	8	14	20	7	13	19	7	13	18	6	11	16
21	8	15	21	8	14	20	7	13	19	7	13	18	6	12	17	6	11	15
22	8	14	20	7	13	19	7	13	18	6	12	17	6	11	16	5	10	15
23	7	13	19	7	13	18	6	12	17	6	11	16	6	11	15	5	10	14
24	7	13	18	6	12	17	6	11	16	6	11	15	5	10	14	5	9	13
25	6	12	17	6	11	16	1	11	15	5	10	14	5	9	14	5	9	13
26	6	11	16	6	11	15	5	10	14	5	9	14	5	9	13	5	9	12
27	6	11	15	5	10	14	5	9	14	5	9	13	5	9	13	4	8	12
28	5	10	14	5	9	14	5	9	13	5	9	13	4	8	12	4	8	11
29	5	9	14	5	9	13	5	9	13	4	8	12	4	8	11	4	8	11
30	5	9	13	5	9	13	4	8	12	4	8	11	4	8	11	4	7	11

It can be seen from the above that the reduction in crop life-duration for unit increase in temperature will (i) increase with increase in base temperatures for crop growth and (ii) decrease with rise in ambient air temperatures. In a crop, its distinct growth phases and hence its life-duration, have specific degree-day requirements to be fulfilled. There is some evidence that though total cumulative day degrees may vary with varieties of a crop, the percentage distribution of the total day-degrees amongst the phases may be nearly the same (Venkataraman et. al. 2005).

Literature on weather relations of crops (Venkataraman and Krishnan, 1992) shows that day temperatures during maturity of soybean, night-temperatures after flowering in wheat and rice, both day and night temperatures during ripening of maize affect crop yields. The concept of Thermoperiodicity (Went, 1957), as mentioned earlier, can thus assist in arriving at a much better understanding of the temperature relations of development of crops. The above phenomenon has brought in the need to derive day and night temperatures from the readily available data of maximum and minimum temperatures (Robertson, 1983). Use of maximum minimum temperatures to (i) estimate the number of hours during which temperatures of a given threshold can be expected (Neild, 1967) and (ii) hourly distribution of temperatures and mean night-time temperatures (Venhkataraman, 2002) have been presented.

Some cold loving plants require to undergo a certain amount of chilling in one or more phases before passing on to the next phase. This phenomenon is called Vernalisation. Also buds of many tree species require to undergo vernalisation before their bud dormancy is broken. Even in the same crop type, the need for vernalisation can be specific two types of varieties. For example, winter wheat requires exposure to temperatures of 0 to 8 °C for several weeks to pass from vegetative to reproductive stage. Spring and Alternate wheat types do not have any low temperature chilling requirements.

2.4.2.2 Nyctoperiod

The duration of the night period, called Nyctoperiod, is another factor controlling crop development. The duration of nyctoperiod to which exposure is required before flowering, varies amongst crops and with the same crop with varieties. Those requiring exposure to long and short nights are respectively called long-night and short-night plants. Those that have no night duration requirements for flowering are the Neutral types. There are a separate group of plants, called the

Intermediate Type (Garner and Allard, 1920), which do not flower when the day lengths are less than 12 hours or more than 14 hours. Again, the night sensitive plants can be divided into the "facultative" and "strict" types. In the former, the effects of nyctoperiod are additive i.e., occurrence of unfavourable nyctoperiods only delay flowering. In the Strict type occurrence of unfavourable nyctoperiods obliterate the effects of favourable nyctoperiods preceding them. Leopold (1964) has drawn attention to a further classification involving (i) long-night plants requiring an induction period of short-nights and (ii) short-night plants requiring long-night induction.

2.4.2.3   Photoperiod – Temperature Interactions

Photo-Thermal units, obtained by multiplying the day-degrees by the photoperiod in the corresponding period are held to be more stable than day-degrees (Nuttonson, 1953). Temperatures, especially night temperatures, can obliterate or modify the influence of one another. For example, all cultivars of Potato behave like night-insensitive types when the night temperatures are in the range of 10 to 14 °C (Went, 1957). A modification of the photoperiodic effects by temperatures in case of cowpea and soybean has been noted (Huxley and Summerfield, 1974). The effects of nyctoperiods and temperatures may be opposing or complementary in nature. When the nyctoperiod is above the optimum requirement of a plant, night-time temperatures would become an operative ecological factor (Salisbury1963). The nature and extent of interactive modifications due to nyctoperiod-night-temperature regimes vary amongst crops.

2.4.3   Crop Yield

Crop yield is the resultant of photosynthetic production and respiratory depletion of drymatter and apportioning of the net accumulated drymatter to the economic-yield component of the crop. Under each of the above, differences exist amongst crops as detailed below.

2.4.3.1 Factors- Requirements, Interactions and Crop Influences

The principal environmental factors concerned with photosynthesis are solar radiation, CO_2 and temperature while respiratory depletion of drymatter is mainly governed by temperature. Compensation and saturation points of light, temperature and CO_2 refer respectively to (i) minimum values of light intensity, temperature and CO_2 levels at which photosynthesis equals respiration values and (ii) values at which photosynthesis reaches the maximum. The value of each of the above parameters is dependent on the level of other parameters. This situation gives rise to the concept of Limiting Factors.

Under same environmental conditions, saturation and compensation points for light, temperature and CO_2 vary amongst crops. For example, Rice is unique in having a very high light compensation point which is also temperature dependent with a Q10 of nearly 2.50 in the temperature range of 15 to 30 °C (Ormrod, 1961).

There are differences amongst crops in efficiency of utilization of solar radiation and CO_2. For example maize and sorghum are most efficient users of solar radiation while soybean and tobacco are the least efficient and average users respectively of solar radiation. Similarly wheat is a more efficient user of CO_2 than rice while maize is the least efficient user of CO_2.

2.4.3.2   Photosynthetic Pathways- C3, C4 and CAM Plants

Crops differ in their method of fixation of CO_2. Those following the Calvin cycle produce a 3-Carbon compound as a precursor and are called the C3 plants. Those following the Hatch-Slack pathway produce a 4-carbon compound as a precursor and are called C4 plants. A very minor group of plants follow the Crassulacean Acid Metabolism and are called CAM plants. Barley, Beans, Beets, Cotton, Groundnut, Oats, Peas, Potato, Rice, Soybean, Spinach, Sunflower and Wheat are C3 plants. Maize, Millets, Sorghum and Sugarcane are C4 plants. Members of Crassulaceae are CAM plants. The C3 pathway is less efficient than the C4 pathway in production of photosynthates. The CAM plants can fix up CO_2 at night. The C3 plants get light-saturated at 2500 to 5000 foot candles. The C4 plants have no light saturation and their photosynthesis is very little affected in the temperature range of 20 to 30 °C. The CO_2 compensation for C3 plants is high and about 5 to 10 times greater than that of C4 plants. Higher temperatures are required by C4 plants compared to C3 plants. Thus, C4 plants are more efficient under hot and bright weather conditions.

2.4.3.3   Respiration

The C3 plants have photorespiration i.e., have high rates of respiration in light. The C4 plants have low respiration rates. There are indications that even amongst C3 and C4 crops the maintenance respiration as a fraction of photosynthesis and its Q_{10} value can vary widely (Bishnoi, 1986).

2.4.3.4   Harvest Index, HI

The ratio of the economic yield of a crop to the total drymatter produced by it during its field-life duration is called the Harvest Index (HI). In field trials for validation of Dynamic Crop-Weather Simulation yields, factual data to compute HI are available for many crops. Generally it is seen that (i) HI is low for pulse crops, maximal for grain crops and average for others and (ii) for any given crop cultivar the values of HI remain nearly the same over a wide range of economic yields and total dry matter production caused by thermal, soil moisture and nutritional stresses. Thus, for realisation of maximal yields, stress-free condition for the crop cultivar is a must.

2.4.3.5   Water Requirements

Venkataraman (1985) has drawn attention to the prevalence of and variations in duration of vegetative lag phase, which is the period from germination to active uptake of water by crops. Water uptake is static in (i) American cotton and sugarcane in the month after sowing/planting (ii) up to flowering in groundnut (iii) up to one third of its life after sowing in safflower (iv) the first two weeks in Hybrid varieties of maize, sorghum and pearl millet and (v) in first 4 weeks in dwarf wheat's and finger millet.

Crop water need in the maturity phase varies amongst crops due to differences amongst crops in physiological control of water uptake in this phase (Hattendorf et. al. 1988; Venkataraman, 1985). For example (i) wheat will have little transpiration at physiological maturity (ii) Hybrid varieties of millets and sorghum have little physiological senescence and their peak water needs continues in the maturity phase (iii) in hybrid maize and groundnut, reduction in water uptake occurs only near harvest (iv) water uptake rapidly declines after flowering in safflower and (v) sugarcane regulates its water uptake during maturity to 50% of the peak value. Local varieties of sorghum however show the ability to curtail water consumption under soil moisture insufficiency (Venkataraman, 1981).

2.5 Agricultural Applications

In light of the above, a basic understanding of micro-climatology, morphology and physiology of crops is necessary to overcome climatic constraints in crop production through agronomic techniques and/or breeding of Ideotypes (Donald, 1968), defined as a plant with an architecture and phasic development suited to a given environment for maximal production. A few examples on the above aspects are set out below.

2.5.1 Rice

The SRI system of rice cultivation is based on an understanding of rice physiology that young rice seedlings retain their potential for formation of tillers if they are transplanted before the start of the 4th phyllochron (Stoop et al, 1992), i.e., before 15 days of age in tropical conditions. In each phyllochron one or more phytomers i.e., set of tiller, leaf and root are produced from the apical meristem and the number of tillers, leaves and roots will depend on the number of phyllochrons completed before flowering (Satyanarayana, 2005). The number of phyllochrons completed in the vegetative phase of rice can vary with variety, season and location and the above may account for areal variations noticed in the above three parameters.

Using the fact of the (i) maturity duration of rice cultivars being 30 plus or minus 5 days under a wide regime of climates, cultivars and cropping systems (Oldeman et. al. 1987) and (ii) need of rice for bright days (Moomaw and Vergara, 1964) and cool nights (Seshu and Caddy, 1984) in the maturity period, cropping systems can be designed to maximize rice yields

2.5.2 Groundnut

The fact that Groundnut is non-thermoperiodic and photo-insensitive with a vegetative duration of only 30 days under a wide range of climates, cultivars and cropping systems, can be used to (i) extend its spatial spread and seasonal cropping period and (ii) maximize its unit area yields.

2.5.3 Sugar Crops

Suitable cultivars of the cane crop in Peninsular India can be planted as close to June as possible to ensures that the leaves are not mature enough in August to act as receptors for the photoperiodic stimulus and thus prevent the occurrence of "arrowing" in October of the same year of planting. Thus, the cane be kept growing for 18 months and also avoid cyclone damage in Eastern coastal regions of India (Venkataraman, 1976). Subjecting the early and late flowering cane varieties in which flower initiation occur in August, to carefully manipulated photoperiod subsequent to floral initiation, can bring about synchronization in flowering of both classes of cane varieties and aid their cross breeding (Vijayasarathy et. al. 1957; Wilsie 1962). The eco-physiology of flowering and seed setting of sugarcane and sugar beet respectively have a great bearing in the location of breeding stations for them.

2.5.4   Grapes

Pruning of grapes in the rainy season in mild-winter areas of Peninsular India helps obtain unit area yields of grapes comparable to other regions in the world.

2.5.5   Soybean

The extension of soybean cultivation in U.S.A. from North Central America to other areas of the Corn Belt, the Missisipi delta and the Atlantic coastal region is due to use of varieties that could withstand higher temperatures than those encountered in favoured habitats.

2.5.6   Potato

The fact that cultivars of potato become photo-insensitive under a night temperature regime of 10-14 °C (Went, 1957) can be used for locating regions for year round production of potatoes (Venkataraman, 1968).

2.5.7   Jute

Use of Capsularis and Olitorious varieties of jute to suit weather conditions helps extend cropping period in West Bengal of India.

2.5.8   Light Utilisation

Crop-weather models on penetration of light in crop canopies can be used to specify ideotypes for maximal use of solar radiation in various regions and periods.

2.6   Applied Agro-Eco Physiology

Ecophysiological response of crops have to be taken account of in many disciplines of applied agrometeorology as enumerated below.

2.6.1   Crop-Climate Classification

In climatological studies the term "Agroclimate" is often used to refer to zones based mainly on rainfall, temperature and evaporative power of air and soil types. The term 'Agro' conveys the impression that such classifications can assist in the examination and/or the planning of cropping patterns. In view of the forgoing, it is clear that classification of climatic zones has to be done on a crop-wise, season-wise basis.

2.6.2   Impact Assessment of Climate Change

Climate change is expected to lead to an increase in (i) rainfall variability (ii) mean and night-time air temperatures (iii) concentration of Carbon Dioxide and (iv) cloudiness. In view of the foregoing. It is clear that impact assessments of climate change on crop yields have to be carried out on a crop-wise, season-wise and region-wise basis adopting a holistic approach. The holistic approach involves (a) assessment of (b) increase in yields due to increased CO_2 (c) reduction in yield on account of reduced (i) field-life duration due to higher temperatures (ii) solar radiation and (iii) increase in maintenance respiration due to increased temperatures and (d) adding the above changes algebraically to get the net change (Venkataraman, 2004).

2.6.3   Avoidance of Climatic and Biotic Stresses

A crop is susceptible to a pest or disease at a certain growth stage only and the pest or disease organism does damage only at a certain development stage. Therefore, once endemic areas and periods therein for a pest or disease are identified, either factually or agroclimatically, the strategy of change in sowing date from the usual will help avoid setbacks due to the concerned organism. The same considerations apply to avoidance of extremes of elements like temperature, wind and rains.

2.6.4   Assessment of Crop Drought

The effects of soil moisture stress arising from rainfall vagaries has to be done on a crop-wise, and soil-wise basis and the stress days weighted suitably depending on the phonologically sensitive stage of the crop to soil moisture deficiency.

2.6.5   Provision of Agromet Advisories

Prophylactic action against both biotic and abiotic stresses will be required only if the crop is in a susceptible growth stage at the time of occurrence of the stresses. Therefore, for meaningful and real-time issue of advisories to farmers, to take advantage of or cope with the forecasted weather, it is necessary to have feed-back information on the prevalent state and stage of various standing crops.

2.6.6   Irrigation

The intervals of irrigation in the crop establishment phase and the timing of the last irrigation of a given crop must be based on its physiological control of water uptake. For example, when crop establishment is slow and takes place in warm weather in water-scarce regions and/or periods, drip irrigation will give the maximum cost: benefit ratio. Again, interval of irrigation for sugarcane can easily be doubled during its maturity period.

2.7   Need for Phytotronic Studies

It is difficult to deduce the physiological make-up of numerous crop cultivars released by crop breeders from field studies. It is easier to screen a cultivar for its physiological response to (i) temperature and photoperiods, singly or in combination, to phasic development (ii) temperature, CO_2, and solar radiation, singly or in combination, in net biomass accumulation (iii) environmental stress on harvest index etc., in climate controlled chambers called Phytotrons. Such studies can be used to determine the cultivar-dependent genetic or photoperiodic coefficients needed for use in dynamic crop-weather simulation models.

References

Bishnoi, O.P. 1986. Solar radiation and productivity in India-I: Potential Productivity. Mausam, 37: 501-506.

Donald, C.M. 1968. The breeding of crop ideotypes. Euphytica, 7: 385-403

Doorenbos, J. and Kasam. A.H. Yield responses to water. FAO Irrigation and Drainage Paper, 24: 179 pp.

Garner, W.W. and Allard, H.A. 1920. Effects of the relative length and day and other factors of the environment on growth and reproduction in plants. Jl. Agric. Res. 18: 553-606.

Geiger, R. 1965. The climate near the ground. Harvard University Press, London. 611 pp.

Hattendorf, M.J.; Redelfs, M.S.; Amos, B.; Stone, L.R. and Gwin, R.E. 1988. Comparative water use characteristics of six row crops. Agronomy Jl., 80: 80-85.

Huxley, R.A. and Summerfield, R.J. 1974. Effects of night temperature and photoperiod on the reproductive ontogeny of cultivars of cowpea and soybean selected for the wet tropics. Plant Sciences Letter, 3: 11-17

Leopold, A.G. 1964. Plant growth and development. Tata McGraw Hill. 466 pp.

Monteith, J.L. 1965. Radiation and crops. Expt. Agric. 1: 241-251.

Moomaw, J.C. and Vergara, R.S. 1964. The environment of tropical rice production. Proc. Symp. Mineral Nutrition of the Rice Plant. International Rice Research Institute, (IRRI). John Hopkuns Univ. Press. Baltimore, Maryland, USA, 3-13 pp.

Nield, R. 1967. Maximum–minimum temperatures as bases for evaluating thermoperiodic response. Mon. Wea. Rev. 95: 583-584.

Nuttonson, J.Y. 1953. Phenology and thermal environment as a means for physiological classification of wheat varieties and for predicting maturity dates of wheat. Pub. Amer. Inst. Crop Ecology Washington.

Oldeman, L.R.; Seshu, D.V. and Caddy, F.B. 1987. Response of rice to weather variables In: "Weather and Rice". Publication IRRI, 5-39 pp.

Ormrod, D.P. 1961. Photosynthesis rate of young rice plants as affected by light intensity and temperature. Agron. Jl. 53: 93-95.

Richards, L.J.; Hagan, R.M. and McCalla, T.M. 1952. Soil temperatures and plant growth. In Soil Physical conditions and Plant Growth. Vol.II. Acad. Press, N.Y. 303-480 pp.

Robertson, G.W. 1983. Ed. Guidelines on crop weather models Pub. World. Met. Organisation. Personal Communication. 202pp.

Ronald, L.S. and Cicerone, R.J. 2002. Photosynthetic allocation in rice plants: Food production or atmospheric Methane?. Proc. Nat. Acad. Sci. USA, 99: 11993-11995.

Salisbury, P.B. 1963. Biological timing and hormone synthesis in flowering of xanthium. Planta, 59: 518-534.

Satyanarayana, A. 2005. System of Rice Intensification-Need of the Hour. Spl. Lecture. 34th Res. and extension advisory council meeting. Acharya N.G. Ranga Agricultural University, Hyderabad, India.

Seshu, D.V. and Cady, F.B. 1984. Response of rice to solar radiation and temperature estimated from international field trials. Crop Science, 24: 649-654.

Stoop, W.; Uphoff, N. and Kasam, A. 2002. A review of agricultural research issues raised by the System of Rice Intensification (SRI) from Madagascar: Opportunities for improving farming systems for resource-poor farmers. Agric. Syst. 71: 249-274.

Swain, D.K.; Hearth. S.; Sake, S. and Das, R.N. 2007. CERES Rice model: Calibration, evaluation and application for solar radiation stress assessment on rice production. Jl. of Agrometeorol. 9: 138-147.

Venkataraman, S. 1968. Climatic considerations in cropping pattern. Proc. Symp. "Cropping Patterns in India". Ind. Counc. Agril. Res. 251-260pp.

Venkataraman, S. 1976. On the climatological aspects of crop management in Sugarcane with special reference to irrigation. SISSTA Sugar Jl. Spl. Issue Irrigation. 1-6 pp.

Venkataraman, S. 1981. Lysimetric observations on moisture accretion for and use by M 35 –1 jowar at Solapur. Jl. Maharashtra Agric. Univ. 6: 36-40.

Venkataraman, S. 1985. Agrometeorology as a link in technology transfer for the stabilization of crop outputs. Jl. Ecol. Environ. Sci. 11: 91-103.

Venkataraman, S. 2002. Tabular aids for computation of derived agrometeorological parameter on a weekly basis. Jl. Agrometeorol. 4: 1-8.

Venkataraman, S. 2004. On possible reduction in yields of grain crops in future climate. Jl. of Agrometeorol. (Spl. Issue), 6: 213-219.

Venkataraman, S. and Krishnan, A. 1992. Crops and Weather. Publication, ICAR. 586 pp.

Venkataraman, S.; Kashyapi, A. and Das, H.P. 2005. Phasic distribution of total heat units in wheat cultivars in India. Mausam, 56: 499-500.

Vijayasaradhy, M.; Narasimhan, R. and Nathan, S.T.S. 1957. Experimental studies on the factors controlling flowering in sugrcane. II. Effect of extra dark treatment in relation to age and month of planting. Ind. Jl. Sugarcane Res. Development, 1:151-163.

Watson, D.J. 1956. Leaf growth in relation to crop yield IN: The Growth of Leaves. Butterworths Sci. Publ.; London. 178-191 pp.

Went, F.W. 1957. Experimental Control of Plant Growth. Chronica Botanica. Ronald Press Co. New York. 343 pp.

Wilsie, C.P. 1962. Crop Adaptation and distribution. Freeman and Co. London. 52-59 pp.

CHAPTER 3

Crop-Climate: Analogues and Zones

In the previous chapter the emphasis was on elucidation of reasons for the often observed differential reactions of crops exposed to same weather situations. In this chapter, the emphasis will be on assessment of the weather requirements for optimal growth, proper development and maximal yields of various crops and the agronomic uses to which such assessments can be put to. Ease and adequate quantum of availability of soil moisture is the most important parameter affecting crop yields. Varietal differences in yields of crops are best expressed under no soil moisture stress conditions. In well managed irrigated agriculture, availability of soil moisture as a crop production factor can be ruled out. So the weather requirements of crops have, *a priori, t*o be worked out for irrigated crops.

3.1   Assessment of Crop-Weather Requirements

The techniques explored for arriving at the optimal weather requirements of crops have in the early stages been conventional agroecological studies (Engledow and Wadhan, 1923). As many crops are seen to be not specifically adapted to the climate of their place of origin or discovery, further refined techniques have been indicated by Went (1957), De Villiers (1966) and Venkataraman (1968). These include (i) a detailed study of the climatology of the regions and seasons in which the crop is giving good yields and (ii) the dynamic approach in which individual phenometric crop attributes, times and duration of crop phases and economic component of crop yield are sought to be related to pertinent, concurrent weather parameters in phytotrons and in multi-location cum multiple sowing date field trails. Developments relating to the above approaches are highlighted below.

3.1.1   Conventional Studies

In these, realising that the plant could itself be used as an integrator of weather effects, attempts were made to use plant attributes, singly or in combination, as indicators of likely yields. The crop attributes were irreversible and could be had only up to the time the plant reached the maturity phase. It was seen that even under irrigation, the final realised yield could go down significantly from the expected level, the causative factor being the weather in the maturity phase. Thus, the statement that wheat production in India is a gamble on temperatures (Howard, 1924) gained as much notice as the cliché that Indian budget is a gamble on the monsoon. The unit area yields of many crops showed wide areal variations in a season and seasonal variations at a location. Also, for any given crop, the rotational crops to be raised in relay cropping were not the same across locations in a given season. The above called for fresh approaches for studying crop-weather relationships.

3.1.2   Global Agroclimatic Analogues

Simple graphs depicting mean monthly values of various meteorological parameters are found in many climatological atlases. In the Climograph the abscissa is used to represent the monthly mean of a climatic parameter and the ordinate the mean monthly value of another parameter and the month number is plotted at the coordinates of monthly means. The points are connected to form a geometric figure. The climographs assist in comparison of locations in which a crop is doing well with reference to other suitable areas and periods with reference to geometric similarity of the climographs. In comparing climographs in the Northern and Southern hemisphere one needs to compare equivalent months, namely January in the north with June in the south and so on. The matching areas and seasons are called Climatic Analogues and are also known as Homoclimes.

Kimbal et al., (1967) has used the climographic technique to assess potential lettuce producing areas in California. Similar data in tabular form has been used by Nuttonson to delineate (a) global agroclimatic analogues for selected crops in specific regions of USA like rice and citrus in continental USA (Nuttonson 1965a & b) and farm crops in North Great Plains region (Nutttonson, 1965c) (b) agroclimatic analogues in USA (i) in terms of ecological crop geography of Germany (Nuttonson, 1949) China (Nuttonson, 1947a) and Ukraine (Nuttonson, 1947b). Papadakis (1970) examined the agricultural potentialities of world climates for five classes of crops, namely winter cereals, mid-season crops (sugar, beets, potatoes), other summer crops (cotton, peanuts, cassava), tree crops (bananas, coconuts), and summer cereals. Boshell and Nield (1975) have used climatic data from tropical tea growing areas to identify five regions in Colombia as analogues with tea producing region of the Cameron Highlands of Malaysia. Nield and Boshell (1976) have used the agroclimatic analogue technique to locate areas suitable for growing pineapple in Colombia and had also graded the locations in terms of closeness of the temperature and rainfall regimes to the ideal one. Agroclimatic analogue technique has been used in introduction of exotic parasites and predators in control of insect pests.

3.1.3   The Dynamic Approach

Agroclimatic analogues are not of much help in introduction of new crops in new areas. Crop breeders are continuously evolving Ideotypes (Donald, 1968) of new crops with a morphological structure and phasic development suited to given environments for maximal production. The emphasis is on locally adapted cultivars for maximal performance instead of on cultivars showing a wider climatic adaptation. The above strategy is logical as any variety that does well under optimum conditions can be expected to have plasticity of reaction to environmental stress. On the other hand cultivars breed for stress conditions are not likely to perform maximally under congenial conditions. Now, the environment for best morphological cum phenological expression of a cultivar may not obtain at the crop breeding station. Thus, to prevent slippage of genetic material that may do well in certain other regions and seasons, it is necessary to test them at many places other than the place of their breeding (Venkataraman and Krishnan, 1992). At a given location, a given cultivar has a narrow sowing date window for optimal performance. So at each location, trials with realistic multiple sowing dates for new cultivars becomes essential. Besides multi-location, multiple sowing date field trials, study of new cultivars in controlled environments is necessary to evaluate the physiological make up of the cultivar relating to thermoperiodicity, photosensitivity etc. The above techniques constitute the Dynamic Approach in crop-weather relationship studies.

Field-life duration, optimal growth, balanced phasic development, total dry matter production and its partitioning into the economic crop part determine the unit area yield of a given crop cultivar. The dynamic approach has led to elucidation of the weather relationships of each of the above yield-determining factors. The applicability and relevance of findings from the dynamic approach in delineation of crop-climate zones is examined below.

3.1.3.1 Field-life Duration

Unit area yield of a cultivar when expressed in terms of its field-life duration as Yield per Day, YPD is conservative across locations in a given season (Swaminathan, 1968; Muchinda and Venkataraman, 1988). Derived agrometeorological parameters, such as Degree Days, Photothermal Units etc., for determining field-life duration of crop cultivars in a given environment are available. The above have led to attempts for climate-based zonation of crops on the basis of feasible crop-life duration. The simplest of such a technique is to determine in terms of calendar periods the times and duration during which the day degree and other requirements of the cultivar are met in various areas and seasons. However, such a procedure will not ensure that the crop will grow and develop normally and realise the potential yield in all delineated growing periods.

3.1.3.2 Growth

As a result of the dynamic approach the cardinal range (minimum, optimum and maximum) of specific weather elements for important growth stages and growth processes like germination, tillering/branching, leaf area development, flowering etc., of many crops had come to be elucidated. These reveal that the optimal requirements of various weather parameters for growth of a given crop vary with its growth phases and for the same growth phases vary amongst crops. The differences in optimal weather requirements amongst cultivars of a crop are one of degree rather than of type. Knowing the optimal weather requirements of a crop for the phases from sowing to physiological maturity can help in assessing, from meteorological data, the feasible times and duration of life-period for the crop in various regions and seasons. The degree-days/photothermal units appropriate to the crop in the delineated field-life period can then be determined to decide on the crop variety best suited for realising the potential yield.

3.1.3.3   Phasic Development

Phenology, which is the study of developmental phases of organisms in relation to calendar dates (Lieth, 1974) has emerged as the most potent factor in the elucidation of weather requirements of crops. Phenological data give the times and duration, D of various phenological phases of crops. The rate of development processes R like tillering is expressed as numbers per day. The reciprocal of D is a measure of R which is useful in determining the base, optimal and maximal value of the parameter under study for the process under examination. Thus, phenological data are useful in determining the cardinal values (minimum, optimum and maximum) of meteorological parameters for various crop growth stages and processes.

Temperature and photoperiod influence the initiation and completion of various specified growth phases. For a given region and season, there is no inter-annual variation in the temporal march of the Photoperiod. Photoperiod also shows lesser areal variations than temperature. Thus, successful introduction of new crops and its improved genotypes is determined by temperature and phenology (Aitken, 1974).

For a given phenological phase, the requirements of temperature and photoperiod for its initiation and completion vary amongst crops and for the same crop with its varieties and is the reason for use of cultivar specific genetic co-efficient in using the phenological component of Dynamic Crop-Weather Simulation models.

A balanced phasic development of a crop cultivar within its life-duration is essential to ensure realisation of the potential yield of the economic crop component. For, example, post-floral photosynthesis significantly contributes to yield of grain crops. It is, therefore, desirable to have as long a maturity period as possible in grain crops. Non-optimal weather conditions in vegetative phase for growth and/or development can be compensated by agronomic techniques. However, no such manipulation is possible in the maturity phase. Again, in rice and groundnut the reproductive and vegetative phases respectively are little influenced by the aerial environment. In soybean the phase from pudding to maturity is insensitive to environment. In other words the duration of the reproductive phase in rice, vegetative phase in groundnut and pod maturation phase in soybean are fixed duration in terms of days. The above features add to the complexity of integrating meteorological relations of growth and development for realising potential crop yields.

3.1.3.4   Total Drymatter Production

The net drymatter accumulated by the crop is the resultant of its production by photosynthesis and its consumption due to growth and maintenance respiration. At existing levels of CO_2, solar radiation is supra-optimal for photosynthesis for clear season irrigated crops. Rate of photosynthesis is little affected over a wide range of temperature for most crops. However, in that range respiration increases two times for every 10 ºC rise of temperature. Therefore, the temperature regime is important in net accumulation of drymatter.

Several studies have shown that the Total Dry Matter, TDM produced by a crop is directly proportional to the amount of Photosynthetically Active Radiation, PAR intercepted by it. The term Radiation Use Efficiency, RUE is used to denote the slope of the relationship between the amount of drymatter produced and quantum of PAR intercepted. Now, PAR is 0.45 times Global Solar Radiation, GSR (Monteith, 1965). So RUE can be expressed in terms of GSR also. From eco-physiological considerations of production and respiratory use of photosynthates, one can, a priori, expect RUE to vary between the crop species but also amongst crops in the same species. Now the quantum of intercepted PAR depends on life-duration of the cultivar from ground shading to physiological maturity and its leaf architecture and the temporal march of solar radiation regime in the crop period. The ground-shading stage is a physical and not a physiological one. Time from sowing to ground shading stage is governed by initial population density and temperature. For a given location and season, the duration from ground coverage to physiological maturity is cultivar dependent. The green revolution had led to quantum jumps in unit area yields of crops. Higher yields of new crop cultivars can come from longer crop-life duration, higher net drymatter production per day or better Harvest index, HI. The new crop cultivars that ushered in the Green Revolution had the same life duration as the older varieties they had replaced. The per day yield of Japonica rice, hybrid maize and dwarf

Mexican wheat's are very much higher than those of indica rice, open pollinated maize and tall indian wheat respectively (Tanaka, 1964; Muchinda and Venkataraman, 1988). Harvest indices for most of the high yielding cultivars are in the range of 0.40 to 0.50. So the betterment of yields due to increase in HI to realise the kind of quantum jump achieved can be ruled out. So any increase in yield by the new varieties must have come through better per day production of net drymatter. So both intercepted PAR and RUE will vary with cultivars. Hence TDM will be influenced by varieties and by the temperature and radiation regimes. YPD of rice cultivars is highest in the Rabi season (Jana and Ghildayal, 1969). So optimal weather conditions are needed for best expression of RUE.

3.1.3.5   Harvest Index (HI)

Factual data of economic yield and concurrent total drymatter production have been reported in field studies conducted for validation and/or calibration of predictions of the above from Dynamic Crop-Weather Simulation models. A preliminary examination of Harvest Indices of crops from such data indicates the following.

For any crop cultivar the HI, at a location and in a given year can remain conservative despite significant decreases in economic yields due to temperature and/or soil moisture stresses. HI can be lower when unit area yields under optimal sowing dates are lower in other years. The HI is seen to be low for pulse and oilseed crops and high for grain crops. Varietal variations in HI are also discernible. Harvest indices of new improved varieties seem to have reached a plateau and further improvements in HI of crops will be a daunting task (Mccloud et. al. 1980).

3.2   Climatic Fertility

From the above, it is readily seen that even for the same crop variety the weather requirements for growth, development and yield are different. Under each of the above heads the weather requirements vary amongst crops and for the same crop with growth stages. Weather parameters show temporal, spatial and year to year in situ variations and the degree of intensity of various weather parameters vary with parameters and time scales under consideration. In view of the above, it is not surprising that for a given crop, even under irrigation, there are favoured areas and periods to realise maximal unit area yields. Example, grapes in mild winter areas of Madhya Maharashtra and Telangana, India. Again, while an area may be less suited to a particular crop it may be more productive for another crop. Example, wheat and sugarcane in North and Peninsular India. The above had led to the concept "Climatic Fertility" (Bernard, 1992) to stress the direct link between variability of climatic parameters and variations in the yield potential of crop cultivars.

3.3   Climatic Zonation of Irrigated Crops

The inadequacy of the concept and maps of agroclimate for agricultural purposes and the need for climate zonation maps on a crop-wise basis and season-wise basis for examining and/or evolving regional cropping patterns has been mentioned earlier. The basic aim of irrigated crop planning is to decide on which crops can be grown where, when, how long, how best and how economically with maximum freedom from pests, diseases and weather hazards like droughts, frosts etc.

To realise the above aim, for eco-physiological reasons, as mentioned in the previous chapter, it is necessary to ascertain for all prospective crops on a season-wise and location wise basis (i) the proper growing period for its maximum possible yield and (ii) other alternate growing periods with indications of yield realisable thereof. The latter will be necessary for fitting of the crop suitably under relay cropping at the place. Once the growing periods for the crop is delineated as above at a network of meteorological stations, it is possible to demarcate climatic zones for the crop with intra-regional homogeneity and intra-regional differences in length of growing periods and hence crop yields thereof.

Crop-wise delineation of climate zones will generate for a given location and season a cafeteria of cropping sequences for relay cropping from which crops and their sequencing can be selected to realise the potential yield of each of the chosen crops. For areas in each zone the same agronomic strategy may be adopted. The crop-climate zones delineated as described above can also be used to determine the cropping sequences for irrigated relay cropping even for virgin areas contemplated for provision of irrigation.

For purposes of zonation of irrigated crops, information/data need to be collected, from the literature on weather relations of crops, for all prospective crops on (a) optimum temperatures (day, night-time or mean temperatures as the case may be) for various growth stages in the vegetative phase (b) optimum solar radiation cum temperature regime in the reproductive phase and (c) Pheno-meteorological models or relationships to determine the duration of various growth phases.

3.3.1   Influence of Varieties on Crop-Climate Zonations

The variations in yield of a given cultivar of a crop used for early, normal and late sown conditions gives a measure of relative yield potential of the cultivar in various delineated crop-climate zones. However, the above discussions show that even for a given crop, its meteorological relationships of growth, phasic development, drymatter production and economic yield will vary with varieties. Thus, any crop zonation of the type described above will be variety-specific. However, influence of varietal variations on crop-climate zonations of a number of crops appear amenable to agroclimatic analyses. Crop varieties may include photo-sensitive and photo-neutral ones. For photo-neutral ones there is some evidence that though total cumulative day degrees may vary with varieties of a crop, the percentage distribution of the total day-degrees amongst the phases may be nearly the same for different varieties (Venkataraman et al 2005). Also, the genetic co-efficient for a cultivar for phenological purposes can be taken to be the same as its female parent (Ritchie and Alagirisamy, 1989). For some photo-sensitive varieties the concept of photo-thermal units appears applicable. For a given location and time-period, inter annual variations in temporal march of photoperiod gets ruled out. The implication is that differences in life-duration (i) amongst photo-sensitive cultivars and (ii) between photo-sensitive and photo-neutral cultivars will be conservative on a climatological basis. The corollary is the fact that relative economic yields of cultivars will be the same for a given location and season and across locations. The above aspects need to be more thoroughly studied.

3.3.2   Advantages of Crop-Climate Zonations

Crop-climate zonations will help (a) in the selection of network of minimum and yet representative stations recording observations on many crops as (i) feedback for weather based agrometeorological services and (ii) ground truth for calibration/validation of satellite-sensed agrometeorological data/information (b) in assessing areal extent of applicability of findings of research stations (c) reduction in number of stations for several detailed agroclimatic analyses on a crop-wise basis (d) decide on the proper selection of crops, their growth periods and rotation suited to local climate for maximal yields in relay cropping and (e) evolve contingency crop planning when start of the season is delayed due to delays in harvest of preceding rainfed crop, temperature vagaries, labour problems etc.

References

Aitken, Y. 1974. Flowering time, climate and genotype. Melbourne University Press, 193 pp.

Bernard. E.A. 1992. "L'intensification de la production agricole par l'agrométéorologie". FAO Agrometeorology Working Papers Series No. 1, FAO, Rome, 35 pp.

Boshell, F. and Nield, R.E. 1975. A computer-statistical procedure to determine agroclimatic analogues for tea production in Colombia. Agric. Meteorol. 15: 221-230.

De Villiers, 1966. Climatic methods applied to problems of introducing new crops. Vol. II. Proc. WMO Seminar Agric. Meteorol. Melbourne.

Donald, C.M. 1968. The breeding of crop ideotypes. Euphytica, 7: 385-403.

Engledow, F.L. and Wadhan, S.M. 1923. Investigations on yield in cereals. Jl. Agric. Sci. 13: 390-439.

Howard, A. 1924. Crop production in India- A critical survey of its problem. Oxford University Press, London. 200 pp.

Jana, R.K. and Ghildayal, B.P. 1969. Growth patterns of the rice plants under varying soil water regimes and atmospheric evaporation demands. II. Riso, 18: 15-24.

Kimball, M.H.; Sims, W.I. and Welch, J.E. 1967. Plant climate analysis for lettuce. Calif. Agric. 21(4): 2-4.

Lieth. H. 1974. Phenology and Seasonality Modeling (Lieth ed.) Springer-Verlag, Berlin, 480 pp.

McCloud, D.F. et al. 1980. Physiological basis for increased yield potential in peanuts. Proc. International Workshop on Groundnuts. ICRISAT, Patancheru, India 125-132 pp.

Monteith, J.L. 1965. Radiation and crops. Exptl. Agric. 1: 241-250.

Muchinda, M.R, and Venkataraman, S. 1988. Influence of temperature and moisture on maturation periods of rain fed maize. Productive Farming, September, 20-25.

Nield, R.E. and Boshell, F. 1976. An agroclimatic procedure and survey of the pineapple production potential of Colombia. Agric. Meteorol. 17: 81-92.

Nuttonson, M.Y. 1947a. Ecological crop geography of China and its agro-climatic analogues in North America. Amer. Inst. Crop Ecology. AICE, Washington, USA. 28 pp.

Nuttonson, M.Y. 1947b. Ecological crop geography of the Ukraine and the Ukranian agroclimatic analogues in North America. International Agroclimatic Series. AICE, Washington, D.C. USA, No.1., 24 pp.

Nuttonson, M.Y. 1949. Ecological crop geography of Germany and its agro-climatic analogues in North America. AICE, Washington, USA, 28 pp.

Nuttonson, M.Y. 1965a. Global agroclimatic analogues for rice regions of the continental United States. AICE, Washington, USA, 9 pp.

Nuttonson, M.Y. 1965b. Agro-climatology and global agroclimatic analogues of the citrus regions of the continental United States. AICE, Washington, USA, 42 pp.

Nuttonson. M.Y. 1965c. Global agroclmatic analogues for the Northern Plains Region of the United States and an outline of its physiography, climate and farm crops. AICE, Washington, D.C., USA. 22 pp.

Papadakis, J. 1970. Agricultural potentialities of world climates. Carboba, Bueos Aires, Argentina, 70 pp.

Ritchie, J.T. and Alagairswamy, G. 1989. Genetic coefficients for CERES models. ICRISAT Bulletin No, 12, 27-29.

Swaminathan, M.S. 1968. Genetic Manipulation of Productivity Per Day. Special lecture, Symposium on "Cropping Patterns in India".

Tanaka, A. 1964. "Examples of Plant Performance". Proc. Symp. Mineral Nutrition of the Rice plant. John Hopkins Univ, Press. 37-49.

Venkataraman, S. 1968. Climatic considerations in cropping pattern. Proc. Symp. "Cropping Patterns in India". Ind. Counc. Agril. Res. Pages 251-260.

Venkataraman, S. and Krishnan, A. 1992. Crops and Weather. Publication Indian Council of Agricultural Research, 586 pp.

Venkataraman, S.; Kashyapi, A. and Das, H.P. 2005. Phasic distribution of total heat units in wheat cultivars in India. Mausam, 56: 499-500.

Went, F.W. 1957. Experimental Control of Plant Growth. Chronica Botanical. Ronald Press Co. New York. 343 pp.

CHAPTER 4

The Concept and Relevance of Radiation Balance in Crop Culture

The Sun is the source of energy for all food production on planet earth. An understanding of the production, receipt, dissipation, and use of solar energy is basic to an understanding of the important aspects of solar radiation in production of crops. The Sun is a huge nuclear fusion reactor, in which Hydrogen atoms are fused in several steps to form Helium and in the process release huge amounts of energy. The fusion process in the Sun, called "quantum tunneling", is an inefficient process for production of energy. However, because of its huge mass, the total energy produced by the Sun is enormous. The Sun emits energy by means of electromagnetic waves. The surface temperature of the Sun is about 5500 °C. So the energy emitted by Sun is in short waves. Solar Radiation is, therefore, also called as Short Wave Radiation.

The radiation energy from the Sun falling on unit area of surface held perpendicular to the rays of the Sun at the mean distance between the Sun and Earth is referred to as the "Solar Constant". Terms like Calorie, Langley, Mega joule, Equivalent Evaporation, Joule and Watt are used as measures of the intensity of the Solar Constant. It is, therefore, necessary to know the following quantitative relationships between the above parameters in terms of the more commonly used term Calorie. The equivalent of above terms with reference to Calorie is:

1. Calorie = 1 Langley; 1 Mega joule per square meter per day $(MJ/m^2/day)$ = 23.9 Calories/cm^2/day; Equivalent Evaporation of

1 mm/day = 58.5 Calories/cm²//day; 1 Joule/cm²/day: (J/cm²/day) = 0.239 Calories/cm²/day; 1 Watt/m², (1W//m²/) = 2.06 Calories/cm²/ day.1 µmol/m²/Sec = 0.263 Calories/cm²/day.

Table 4.1 The "Standard Weeks" calendar

Week No.	Month	Dates	Week No.	Month	Dates
	January	1-7	27	July	2-8
2		8-14	28		9-15
3		15-21	29		16-22
4		22-28	30		23-29
5	Jan/Feb.	29-4	31	Jul/Aug	30-5
6	February	5-11	32	August	6-12
7		12-18	33		13-19
8		19-25	34		20-26
9	Feb/Mar	26-4*	35	Aug/Sept.	27-2
10	March	5-11	36	September	3-9
11		12-18	37		10-16
12		19-25	38		17-23
13	Mar/Apr	26-1	39		24-30
14	April	2-8	40	October	1-7
15		9-15	41		8-14
16		16-22	42		15-21
17		23-29	43		22-28
18	Apr/May	30-6	44	Oct/Nov	29-4
19	May	7-13	45	November	5-11
20		14-20	46		12-18
21		21-27	47		19-25
22	May/June	28-3	48	Nov/Dec	26-2
23	June	4-10	49	December	3-9
24		11-17	50		10-16
25		18-24	51		17-23
26	June/July	25-1	52		24-31#

* In Leap Year the week No. 9 will be 26 February to 4 March i.e., 8 days.

\# Last week No. 52 will have 8 days, 24 to 31 December.

4.1   Extraterrestrial Radiation

The solar radiation received by a horizontal surface outside the atmosphere is referred to as "Extraterrestrial Radiation", Ra. The instantaneous intensity of solar radiation on such a horizontal surface is determined by the angle between, the direction of the rays of the Sun and the normal to the surface. This angle will vary with the day, will be different in different latitudes for the same hour and will also be different in different seasons. The value of daily Ra over a given location is thus a function of Latitude (φ), Declination of the Sun (δ), Hour Angle of the sun (h) and the Radial Distance of the earth(r). Thus, Ra can be computed from Solar constant, S from the equation below:

$$Ra = S[(\text{Sin } \varphi, \text{Sin } \delta + \cos \varphi \cdot \cos \delta \cdot \cos h)]/r^2$$

The normal value of S accepted for calculating Ra is 1.94 Calorie/ cm²/minute. However, a value of 1.96 Calories/cm²/minute has been accepted and used by Allen et. al. (1998) to compute monthly values of Ra, in MJ//m² centered around the calendar months, for latitudes 0 to 70 degrees North and 0 to70 degrees South. Time-unit accepted for agrometeorological work is the "Standard Week". Therefore, weekly values of Ra, with S value of 1.96 Calories/cm²/min for 52 Standard Weeks of the calendar year, (Table 1) for 0 to 70 degrees North and 0 to 70 degrees South at 5 degree latitude intervals are given in Table 4.2 A and 4.2 B respectively.

4.2   Atmospheric Modification of Solar Radiation

On passing through a cloudless atmosphere, the solar radiation gets (i) scattered by molecules of air smaller than the wavelength of light (ii) scattered and diffusely reflected by particles of size equal to or greater than the wavelength of light (iii) selectively absorbed by (a) Ozone of wavelengths in the Ultraviolet range (b) molecules of atmospheric gases of X-Rays and (c) Carbon Dioxide and water vapour of wavelengths between 0.85 and 1.30 µ. The earth's atmosphere thus provides a window, which enables us to get solar radiation mainly in the visible and infrared wavelengths. On its passage through the atmosphere, solar rays do not heat the atmosphere. The dissipations in the atmosphere due to scattering, diffuse reflection and absorption in a cloudless atmosphere, as mentioned at the outset, generally account for 25% depletion of Ra. When clouds are present, the solar radiation is scattered, diffusely reflected and absorbed by the cloud mass. Depending on cloud type, namely high, medium or low, 25 to 70% of Extraterrestrial Radiation, Ra reaches the earth's surface mainly as Diffuse Sky Radiation. It is due to the above that Solar Radiation is also known as Global Radiation to indicate that it is the sum of the direct Short-wave Radiation from the Sun and Diffuse Sky Radiation from all upward angles of the sunlit sky.

Table 4.2 A Weekly extra-terrestrial radiation, Cal./Cm²/Day northern hemispheric latitudes

Std. Week	0	5	10	15	20	25	30	35	40	45	50	55	60	65	70
1	863	808	755	693	629	559	492	418	344	270	198	129	67	19	0
2	865	813	758	700	638	569	502	428	353	280	208	136	74	21	0
3	868	820	770	712	650	590	516	445	371	299	227	155	48	33	7
4	877	829	781	727	667	605	538	468	397	325	253	182	115	55	19
5	882	839	793	743	686	624	562	492	423	351	280	208	139	74	31
6	889	851	810	762	710	650	590	523	457	385	313	241	170	100	45
7	896	863	825	781	731	676	617	552	488	418	344	272	198	127	62
8	899	872	839	801	758	705	650	590	528	461	390	323	249	175	107
9	901	879	853	820	781	734	683	629	569	504	537	371	299	225	155
10	903	887	868	839	805	762	717	667	609	547	485	418	370	275	203
11	906	896	882	858	829	791	751	705	650	599	531	466	397	323	253
12	901	896	887	872	848	820	781	743	695	643	585	526	461	394	323
13	894	896	891	879	860	834	801	767	722	674	621	562	502	435	368
14	889	896	896	891	877	858	832	801	762	719	672	617	562	499	437
15	882	896	903	906	899	884	868	844	810	774	731	683	626	569	511

Table 4.2 A *Contd....*

51

Std. Week	0	5	10	15	20	25	30	35	40	45	50	55	60	65	70
16	872	894	906	911	908	903	887	870	844	810	774	731	683	636	585
17	863	887	903	913	915	913	903	887	865	837	803	765	724	679	633
18	853	879	901	915	927	930	922	913	899	875	849	817	781	746	717
19	841	872	899	918	935	942	944	939	935	913	891	868	839	813	793
20	829	865	899	920	939	951	958	958	954	942	925	906	882	863	853
21	820	863	896	920	939	954	963	968	963	954	961	923	901	887	887
22	815	856	891	918	942	958	970	977	975	968	958	944	930	920	930
23	805	849	887	915	944	963	977	987	987	985	977	968	958	954	973
24	798	844	884	915	944	968	985	997	1001	1001	997	992	985	987	1016
25	800	846	887	915	944	966	975	992	997	997	989	982	975	975	1001
26	803	849	887	915	942	963	954	989	989	989	982	975	966	963	985

Table 4.2 A *Contd...*

52

Std. Week	0	5	10	15	20	25	30	35	40	45	50	55	60	65	70
27	805	849	887	915	942	963	961	985	985	985	973	966	951	951	966
28	810	851	887	915	939	958	968	977	975	970	963	954	942	935	949
29	815	853	887	915	937	954	963	966	963	958	944	927	925	901	908
30	825	858	889	913	932	946	951	951	944	937	920	901	882	863	860
31	832	863	891	913	927	937	939	935	925	913	891	868	844	820	805
32	844	870	894	911	920	925	925	915	901	882	858	827	796	765	739
33	853	877	896	911	915	915	908	896	877	853	822	786	748	710	669
34	863	882	894	901	901	894	882	863	837	805	777	729	683	638	590
35	872	884	891	891	887	872	853	829	796	760	729	672	619	569	514
36	877	887	889	884	877	858	834	805	767	729	691	631	576	521	461
37	887	891	887	875	860	837	805	770	729	681	631	571	511	449	378
38	889	887	879	860	839	810	774	734	688	633	578	516	454	387	315
39	891	882	868	846	820	794	746	700	650	583	535	471	404	339	268
40	891	877	868	839	798	758	715	664	617	550	488	421	351	284	213
41	893	870	844	810	772	724	676	619	562	495	430	359	289	217	146

Table 4.2 A Contd…

53

Std. Week	0	5	10	15	20	25	30	35	40	45	50	55	60	65	70
42	891	863	832	793	750	698	648	585	523	456	387	315	246	175	105
43	887	853	820	776	729	676	614	557	495	425	356	284	215	146	84
44	882	846	805	760	710	653	595	530	464	394	323	253	184	117	60
45	875	839	791	739	686	624	562	495	428	356	284	213	146	81	31
46	868	825	774	719	662	597	531	461	390	318	244	175	105	45	2
47	863	817	767	710	652	585	519	446	375	303	232	160	93	36	2
48	860	810	760	703	641	576	507	435	361	289	317	148	84	29	1
49	856	805	753	695	629	562	492	421	347	273	201	134	69	21	1
50	851	800	743	681	617	547	478	404	330	256	184	117	57	10	0
51	853	801	746	683	619	550	480	409	332	261	189	119	60	12	0
52	856	803	748	686	621	552	485	411	337	263	191	122	60	14	0

Table 4.2 B Weekly extra-terrestrial radiation Cal./CM²/Day southern latitudes

Std. Week	0	5	10	15	20	25	30	35	40	45	50	55	60	65	70
1	863	901	944	977	1004	1025	1037	1047	1049	1044	1037	1025	1013	1006	1023
2	867	906	944	977	1001	1020	1032	1040	1040	1035	1025	1011	999	987	1001
3	868	908	944	975	994	1009	1016	1021	1025	1016	1004	987	970	954	958
4	877	911	944	968	985	999	1001	1004	994	980	963	944	913	894	911
5	882	913	942	963	977	987	987	983	973	956	935	911	882	853	837
6	889	915	939	958	968	970	968	961	942	920	891	863	827	791	760
7	896	920	939	951	956	954	946	935	911	882	851	813	772	727	689
8	899	918	930	935	935	927	913	896	868	834	796	753	707	657	637
9	901	915	920	920	915	901	882	852	825	786	743	695	643	585	531
10	903	911	911	906	896	875	851	820	781	739	691	636	578	516	454
11	906	908	903	891	875	849	820	781	739	691	636	576	514	447	378
12	901	896	887	872	849	820	786	743	698	645	590	526	461	394	325
13	894	887	872	851	822	791	751	705	657	600	549	476	411	344	273
14	889	874	856	829	786	758	712	662	609	550	495	421	354	284	315
15	882	860	834	801	760	717	667	612	555	488	423	351	284	213	143

Table 4.2B *Contd....*

Std. Week	0	5	10	15	20	25	30	35	40	45	50	55	60	65	70
16	872	849	817	777	734	686	633	574	514	447	380	311	239	165	103
17	863	831	801	758	710	660	605	543	480	413	354	277	208	131	79
18	853	820	779	734	683	629	571	509	445	375	306	239	170	96	53
19	839	803	758	707	655	597	535	471	404	335	265	196	131	65	24
20	829	787	741	686	631	571	507	442	373	303	234	165	100	43	5
21	822	772	722	676	619	557	492	428	356	289	217	151	88	36	2
22	815	765	715	664	605	543	476	411	341	272	203	136	77	26	2
23	805	758	705	653	590	526	459	394	323	256	186	122	38	17	2
24	798	748	695	638	576	509	442	375	306	237	170	105	62	7	0
25	800	751	700	647	581	514	447	380	311	241	175	110	53	10	0
26	803	753	703	645	583	519	454	387	315	249	182	115	57	13	0

Table 4.2 B *Contd....*

56

Std. Week	0	5	10	15	20	25	30	35	40	45	50	55	60	65	70
27	805	758	707	650	588	523	459	392	323	253	186	122	62	17	0
28	810	760	710	655	595	531	466	399	330	261	194	129	67	19	0
29	815	770	722	667	609	547	483	416	347	280	213	146	84	33	7
30	822	779	734	681	624	564	502	437	371	301	234	167	103	48	17
31	834	793	751	700	645	590	528	466	399	332	263	196	131	69	29
32	844	808	767	719	672	616	557	495	433	363	294	213	160	93	45
33	853	822	786	741	695	643	585	526	464	394	327	258	189	119	60
34	863	834	805	765	722	674	621	564	504	440	373	303	237	165	103
35	870	849	820	784	746	700	653	597	540	478	413	349	275	205	143
36	877	863	839	810	774	734	688	638	583	523	461	394	325	256	189
37	887	877	860	839	808	774	731	686	633	576	516	452	385	313	241
38	889	887	875	858	834	803	767	724	676	624	566	504	440	371	301
39	891	891	887	877	858	833	801	765	727	674	621	564	502	437	371
40	891	899	899	894	879	860	839	803	765	722	674	621	562	502	440
41	894	908	913	913	908	891	872	846	815	777	734	681	631	576	516

Table 4.2 B Contd....

Std. Week	0	5	10	15	20	25	30	35	40	45	50	55	60	65	70
42	891	911	923	930	930	920	906	887	858	825	786	743	703	648	595
43	887	911	925	943	942	937	927	913	889	860	827	789	762	705	662
44	882	908	930	944	954	954	948	939	920	896	868	834	798	762	727
45	875	906	935	954	968	975	977	973	958	942	920	891	863	834	810
46	868	906	939	966	985	997	1004	1004	999	987	970	949	925	903	891
47	863	903	939	968	992	1006	1016	1016	1016	1006	992	975	956	939	939
48	860	901	939	973	994	1011	1023	1025	1028	1021	1009	994	977	968	973
49	856	901	939	973	999	1018	1030	1035	1040	1035	1028	1016	1004	997	1011
50	851	899	941	977	1004	1028	1047	1054	1061	1059	1056	1047	1040	1042	1068
51	853	901	944	977	1004	1028	1044	1054	1059	1057	1053	1045	1035	1035	1068
52	856	901	944	977	1004	1028	1042	1054	1056	1054	1049	1042	1030	1028	1052

4.3   Relative Solar Radiation

The term Relative Short Wave Radiation is used to denote the ratio of Solar Radiation, Rs that actually reaches the earth's surface in a given period to the Solar Radiation that would have been received under cloudless sky for the same place and period, Rso. The ratio of Rs/Rso gives a measure of the cloudiness of the atmosphere. The ratio of actual hours of bright sunshine, n to the feasible hours of bright sunshine at a location and time, N is known as Relative Sunshine duration. The ratio n/N also gives a measure of cloudiness of the atmosphere. For a given location and time Rs/Rso can be taken as equal to n/N. As will be detailed later, the cloudiness factor is of importance in calibration of observed data on solar radiation, Rs in terms of Ra. Considerable data on "n" and Rs are available for a very large number of stations and many stations respectively. Data on "N" are available for monthly periods. In view of the importance of information on "N" in agroclimatic studies on solar radiation, weekly values of N for the standard weeks for "0 to 70 degrees North, and 0 to 70 degrees South latitudes at 5 degree" intervals are given at Tables 4.3 A and 4.3 B respectively.

4.4   Back Radiation

4.4.1   Reflection

Solar Radiation impinging on a surface of the earth is reflected back. The reflection coefficient, known also as Albedo, varies with the type of surface. For the same surface, albedo increases with decreasing elevation of Sun. However, the energy reflected at low elevations of the Sun is very small and determination of albedo values at noon is considered well enough for all practical purposes. Value of Albedo ranges from 0.05 for water to 0.95 for freshly fallen Snow. Ice has a lower albedo than snow. Light coloured soils have higher albedo than dark coloured ones and moist soils have much lower albedo compared to dry soils. Some albedo values for soils reported by Venkataraman and Krishnan, (1992) are: Dry, Black soil: 0.14; Moist, Black soil : 0.08; Dry Gray soil: 0.25-0.30; Moist ploughed field: 0.14-0.17; River Sand: 0.43; White Sand: 0.34-0.40; Loamy Desert soils: 0.30; Desert Yellow Sand: 0.35; The albedo for a short reference green crop cover is taken as 0.23 (Allen et. al., 1998). However, for tall crops and trees the albedo values are lower than for a short crop on account of internal reflections and absorption of solar radiation.

4.4.2   Surface Heating

The heat source for the atmosphere is the back radiation from the surfaces of the earth. This is the reason for the drop in temperature with height in the atmosphere. Factors like penetration and storage of solar radiation, evaporative use and exchange of heat between the surface and the overlying air, determine the fraction of non-reflected solar radiation that is available for surface heating. The fraction of non-reflected solar radiation used up in each of the above processes depends on the type of surface and significantly varies amongst them. In soils, depth of penetration of solar radiation increases with coarseness of soils. Even in coarse soils, much of the heat absorbed during the day tends to be released during the night. However, the heat exchange during the day affects the surface temperatures. Thus, in soils exposed to same solar radiation, while the surface minimum temperature tends to be the same in all soils, there are considerable differences in maximum surface temperatures. The mean radiative temperature T_s of various soils will, therefore, be different. Even for the same soil and radiation receipt, Ts is influenced by the moistness of the soil surface and will be lower for moist soils on account of use of energy for evaporation of moisture. Again, the darker the soil, the more will be T_s. The amount of energy lost or gained by the soil due to its heating or cooling is small compared to net short wave radiation but should be taken into account in computation of radiation balance of crop surfaces, In shallow water surfaces of 10 cm depth, the temperatures are isothermal with depth and up to 35 cm the T_s is the same. With deeper water bodies the daily range in T_s varies in an inverse proportion to the 4h root of depth. Deep water bodies can absorb about 90% of the solar radiation. In case of crop surfaces, the crop material taking part in the disposal and absorption of solar radiation have very small thermal capacities. Thus, T_s of a cropped surface is principally governed by sensible and latent heat fluxes. For an actively transpiring crop fully covering the ground, mean T_s will be equal to the mean temperature of the air (Linacre, 1968). When crop cover is incomplete the mean T_s will lie between that of a full-grown crop and that of bare soil on which it stands. In case of snow, the extent of extinction of non-reflected solar radiation depends on density of snow. Ice permits greater penetration of solar radiation and its T_s is much greater than that over snow. It is the differences in T_s exposed to the same solar radiation that give rise to differences in microclimates, namely the profiles of temperature and humidity, over various surfaces and inside cropped surfaces.

4.5   Long Wave Radiation

The solar radiation absorbed by a surface is converted into heat energy. The heated surface emits radiative energy, called Terrestrial Radiation, at wavelengths much lower than that of solar radiation, namely 0.5 to 80 μ with maximal intensity at 10 μ. Thus, Terrestrial Radiation is also known as Long Wave Radiation. The emitted radiation constitutes the upward flux of long wave radiation and is designated as Rlup. Part of Rlup is absorbed by the Carbon Dioxide and water vapour in the atmosphere and rest is lost to space. The absorption of Rlup by the atmosphere increases its temperature and as a result the atmosphere also emits long-wave radiation Rl down. Thus, the earth both emits and receives long wave radiation. The difference between Rlup and Rl down is called Net Long Wave Radiation, Rln.   Rlup is always greater than that of Rl down. Therefore, Rln represents a loss of energy

All surfaces on earth are good absorbers and emitters of long wave radiation and can be treated as true black bodies. The albedo and emissivity of surfaces for long wave radiation are independent of the nature of the surface and have values of 0.03 and 0.97 respectively. In view of the foregoing, it can be surmised that over large fetches of natural surfaces like oceans, snow and forests the quantum of long wave radiation from the atmosphere can differ significantly.

4.6   Net Radiation

The difference between the non-reflected short wave radiation and net long wave radiation is called Net Radiation Rn. Net Radiation is generally positive during day and negative during night. For, a 24 hour period Rn is positive. Net Radiation is an important parameter affecting water needs of crops and hence in meteorological estimation of crop water requirements.

4.7 Measurement of Components of Net Radiation

4.7.1 Solar Radiation

Solar radiation is measured by instruments in which the sensor is a Thermopile, containing a number of thermocouple junctions in series. The thermopile sensor is suitably covered by a glass or quartz dome that transmits only short wave radiation. The hot junctions of the thermopile are painted black while the cold junctions are painted white. Temperature differences arise out of the absorption and reflection of solar radiation by the two sets of junctions and the difference gives a measure of solar radiation intensity. Alternately the cold junctions may also be painted black but attached to a massive metallic block set flush with the sensor surface but protected from radiation such that its temperature changes only very slowly. The Eppley Pyranometer belonging to the former category and the Moll Gorczynski Pyranometer belonging to the latter category are the most widely used instruments. For many agricultural applications, a cheaper type of Pyranometer is desirable. For this, one could use the Bimetallic Actinograph, the Spherical Pyranometer or the Solar Cell Radiometer. The last two have several disadvantages and inaccuracies in measurement and hence not recommended for climatological data build-up of solar radiation. The Bimetallic Actinograph consists of three strips of bimetals in which the central one is coated black while the others are coated white. The white strips are fixed at one end and support the black strip at the other. The bimetallic strips are covered by a hemispherical glass dome and oriented in the east-west direction while in operation. The differences in temperature between the strips on account of differential absorption of solar radiation causes differential bending of the strips and leads to movement of the free end of the black strips. This movement is proportional to the solar radiation received and is recorded by mechanical linkage wound around a clock drum. The area traced by the curve is planimetered to obtain daily total solar radiation.

4.7.2   Photosynthetically Active Radiation

Measurement of Photosynthetically Active Radiation (PAR) intensity, has assumed importance of late. PAR is given by the total energy contained between the wavelengths of 0.4 and 0.7 µ. For this, the normal dome of the Pyranometer is replaced by specially made optical domes that filter out the wavelengths outside the above range. The use of such domes, however, changes the calibration factor of the instrument for measurement of Total Global Solar Radiation. Thus, Pyranometers for measurement of PAR must be separately and suitably calibrated.

4.7.3   Diffuse Short Wave Radiation

The diffuse short wave radiation from the sunlit sky is measured by screening the surface of the Pyranometer by means of a shading ring. The shading ring is mounted on an adjustable bar to ensure that with periodic adjustments in the tilting of the ring, the surface of the Pyranometer will remain shaded from sunrise to sunset. Corrections for the diffuse radiation intercepted by the ring have to be and are made.

4.7.4   Reflected Solar Radiation

The reflected short wave radiation can be measured by mounting the Pyranometer such that the sensing surface is horizontal and faces the surface under consideration. Use of double pyranometers such that one faces the surface while the other is exposed to the atmosphere is preferable to compute hourly values of albedo of the surface from the generated data. The double Pyranometer must be exposed over an adequate fetch and areal spread of the surface under consideration.

4.7.5   Soil Heat Flux

The quantum of heat flowing into and out of the soil is measured by the use of soil heat flux plates, which are minute net pyrradiometer with the thermopile junctions located in the upper and lower faces. The plates have known and specific thermal conductivity. For measurements, the thermal conductivity of the selected plates must nearly be the same as that of the soil. The plates should not be placed at depths greater than 2 cm from the surface and should be suitably secured to prevent dislocation by earthworms etc.

4.7.6   Atmospheric Long Wave Radiation

It is customary to arrive at Atmospheric Long Wave Radiation by measuring both, the incoming long wave and short wave radiation by means of a hemispherical Pyrradiometer and subtracting from the measured values the value of the short wave radiation as measured by a similarly exposed Pyranometer. The Pyrradiometer for this measurement consists of a matt-black receiving surface enclosed in a polythene dome that is transparent to both short and long wave radiation. The temperature of the receiving surface measured in this device gives a measure of the total radiation received. In other words, by replacing the glass dome of a Pyranometer with a suitable polythene dome, the Pyranometer can be converted into a Pyrradiometer.

4.7.7   Net All Wave Radiation

This entity is measured by Net Pyrradiometers, in which the sensor may consist of a horizontal flat surface suitably and similarly blackened on both sides or of two blackened sensing elements placed back to back. The sensing element is shielded by covering both surfaces by a polythene dome. If the dome is thin it has to be kept that by air pressure. For thicker domes this is not necessary. The instrument is freely exposed to the air a little distance above the surface. The temperature difference of the upper and lower surfaces of the sensing elements give a measure of the net flux of energy and is measured by a thermopile. In this, the temperatures of the upper and lower surfaces can also be measured separately, these, when corrected for the long wave radiation emitted from the instrument to the surface can give the incoming and outgoing fluxes of net radiation. For this, a measure of the temperature of the instrument by thermocouple thermometry with the reference junction located at a depth of one meter in the soil, where the diurnal variation in temperature is almost non-existent, is required.

4.7.8   Thermal Back Radiation

For this radiative skin temperature of the surface is required. This is done by Infra-red thermometers that measures the upward flux of radiation in the range of 8.5 μ to 11 μ (that constitute the atmospheric window) are used. Portable handy models that indicate the determined fluxes in terms of T_s are also available.

4.8   Estimation of Net Radiation and Its Components

Measurement of Net Radiation and its components require the use of highly sophisticated instruments and scientific manpower. Hence, their availability is restricted to few locations. However, net radiation and/or its associated data arc needed for agroclimatological work relating to irrigation scheduling, improvement of crop efficiency in utilization of solar energy and in organizing mitigation measures against the expected change in agromet parameters in the near future on account of the envisaged climate change. Therefore, transposing of available data on net radiation and its components to as many locations as possible becomes important. In view of our advanced understanding of energy balance relating to cropped surfaces, it is possible to do so through other more widely observed meteorological parameters. Procedures for effecting the transposing of data on Net Radiation and its components are detailed below element-wise.

4.8.1   Global Solar Radiation

Relating the ratio of solar radiation expected to be received on clear days Ro to the value of extraterrestrial radiation Ra for that place and time to relative sunshine duration, n/N, had been suggested and tried. However, the ratio of Ro to Ra is dependent on the perceptible water content PPW of the atmosphere. Data on PPW are scarce and attempts to relate PPW to vapour pressure of air at ground level have proved to be not fruitful. Suggestions have been made to relate ratios of actual solar radiation received Rs to Ra to relative sunshine by the following relationship, namely:

$$Rs/Ra = (a + b * n/N)$$

In the above widely followed relationship, it has been realized that even on very clear days the value of n/N will only be 0.90 instead of 1.0. While values of n/N are readily available for weekly periods, data on Rs can be had only for monthly periods. Further values of "a" and "b" show both temporal and spatial variations. To overcome this problem Gangopadhyaya et al., (1970) have advocated (i) the joining of radiation stations in the net work by straight lines and erecting perpendicular bisectors on each of the lines to form a series of polygons which contain only one station in a polygon and (ii) that the constants a and b determined for the radiation station to be applicable to all sunshine recording stations in the area of the polygon. Currently concurrent monthly data of Rs and n are available for a very large number of stations. Therefore, direct interpolation of Rs values between stations may be possible failing which the constants a and b determined for each of the radiation stations may be interpolated.

4.8.2 Long Wave Radiation from the Atmosphere Rlw↓

The above component is by far the most difficult of estimation. In India, the clear sky atmospheric radiation, Rlwo , computed from the formula of Idso and Jackson (1969) in terms of screen level air temperature (Venkataraman and Krishnamurthy, 1972) are seen to agree with observed values of Mani et al., (1965) with an accuracy of plus or minus 5%. However, in view of the large value of Rlwo↓, compared to Rn, an accuracy of 5% will not be good enough for computing Net Radiation. Further the influence of clouds has also to be taken in to account on many days to get at actual values of Rlwo↓. Accounting for effects of clouds in arriving at Rlw↓ is more complex than that for solar radiation. The Pyrgeometer is the instrument used to measure net long wave radiation. Kale (1951) has demonstrated the feasibility of measuring Rlw↓ during daytime also by suitable modifications to the conventional Pyrgeometer. However, routine observations of Rlw↓ are done during the night-time only. In India, Pyrgeometric data on ratio of net long wave to back radiation recorded at 4 specified hours at night are available for a fairly, large number of station-years. Venkataraman (1977) has shown that for climatological estimation of Rlw↓, such ratios would apply for the day as a whole. Now, all radiation stations are also Pyrgeometric stations. Therefore, at the radiation stations, multiplying the climatological monthly ratios of net to back radiation by BT_s^4, where T_s in mean air temperature in degrees absolute and B is the classical Stefan-Boltzmann constant will give Rlw↓.

Since data on Rlw↓ are scarce, Allen et. al. (1998) advocate the following empirical formulation, for computing the net long wave flux, namely:

$$Rnl = B/2\left[\left\{T_{max}\ K^4 + T_{min}\ K^4\right\}(a_h - b_h\ \sqrt{e_d})\ \left(a_c \times (Rs/Rso) - b_c\right)\right]$$

In the above Rnl is the net long wave radiation; T_{max}., K and T Min., K are Maximum and Minimum Absolute Temperatures; e_d is Vapour Pressure of Air; Rs/Rso is Relative Shortwave Radiation; the constants a_h and b_h relate to the humidity factor and ac and b_c relate to the cloudiness factor. For values of Rnl in $MJ/m_2/day$ and e_d in kilo Pascal the values of 0.34 for a_h. 0.14 for b_h, 1.35 for a_c and 0.35 for b_c have been suggested by Allen et. al. (1998). However, it is not necessary to calculate back radiation separately at maximum and minimum temperatures. Back radiation can be calculated at mean absolute air temperature, Ta K. Also Rs/Ro can be replaced by n/N. Thus, for values of e_d in mm and Rnl in calories/cm²/day the above equation may be written as follows, namely:

$$Rnl = B \{Ta K^4\} (0.34-0.0511 \sqrt{e_d}) [(1.35 n/N) - 0.35]$$

On the basis of observed values of net long-wave radiation, calibrated constants in place of the suggested 0.34, 0.0511, 1.35 and 0.35 can be and needs to be worked out.

4.8.2.1 Vapour Pressure and Thermal Back Radiation

For atmospheric Vapour Pressure, units such as Millimeters (mm), millibars (mb) or kilo Pascal (kPa) are used. The conversion factors for above units are: 1 Mb = 0.10 kPa; 1 mm = 0.1333 kPa; 1 kPa = 7.50 mm. Tables for computation of vapour pressure and thermal back radiation values from routine meteorological data, for use in above equation can be had from Tables 4.4 and 4.5 respectively.

4.8.3 Reflected Short Wave Radiation RS↑ and Long wave Radiation RLW↑

The fluxes of short and long wave radiation emanating from a surface will, as detailed earlier, vary with the type of surface. For agroclimatological studies, a well-watered, ground-covering, green, vegetative crop may be taken as a reference surface for computing Rs↑ and Rlw↑. For such a surface Rs↑ may be taken as 23% of incoming solar radiation (Allen et. al; 1998). As mentioned above, for the above crop surface, T_s will be equal to the mean air temperature. The surface will also reflect 3% of the incoming long-wave atmospheric radiation. Therefore, for the above surface RIw↑ will be BT_s^4 plus 3% of RIw↓ where T is mean air temperature in degrees absolute, B is the Stefan-Boltzman Constant and RIw↓ is as estimated above.

4.8.4   Photosynthetically Active Radiation (PAR)

Data on PAR are more scarce than solar radiation. However, studies indicate that PAR may be taken as 45% of solar radiation (Monteith, 1965). The above relationship needs to be verified and/or calibrated for other places with concurrent data on PAR and solar radiation.

4.9.   Agricultural Implications of net Radiation and/or its Components

4.9.1   Net Radiation and Water Requirements of Crops

The concept of Potential Crop Transpiration, PCT (Smith et. al; 1992) provides a very useful tool for (i) determining peak water needs of crops (ii) for unification of apparently diverse data on water needs of crops (iii) pin-pointing crop and soil factors that influence crop water needs and (iv) agrometeorological scheduling of irrigation for a given crop and location. Net Radiation is a vital component in determination of PCT. The concept and use of PCT will be dealt with in some detail in the portion on meteorological determination of crop water needs in the next Chapter on Agrometeorology of Crop Water Usage.

4.9.2   Crop and Radiation-based Relationships in Photosynthesis

Crop yield is ultimately a question of photosynthetic production and respiratory consumption of drymatter and apportioning of the net accumulated drymatter to the economic yield-component of the crop. Photosynthesis is better when sun's rays are slanted. The more the photoperiod the more is the photosynthesis. Because of the above, there are differences in the yields of crops across latitudes with better yields at higher latitudes. The principal environmental factors concerned with photosynthesis are solar radiation, CO_2 and temperature while respiratory depletion is mainly governed by temperature. In environmental relations of photosynthesis, the term light intensity instead of solar radiation is used. Other terms used are Foot Candle, Saturation light Intensity, Light Compensation Point and CO_2 Compensation point.

In terms of energy, one calorie/cm^2/minute is equal to 7000 foot candles. Intensity of sunlight at noon in summer in tropics will be about 10000 foot candles. Saturation Light Intensity is the minimum light-intensity at which rate of photosynthesis reaches the maximum value. Temperature, Light and CO_2 Compensation points refer respectively to minimum values of temperature, light intensity and CO_2 level at which the respiration rate equals the photosynthetic rate. The value of each of the above parameters relating to temperature, light intensity and temperature is dependent on and vary with the level of other parameters and gives rise to the concept of Limiting Factors (Blackman, 1905) in determining rates of photosynthesis in plants.

The response of crops to a given environmental factor is affected by growth stage and physiology of crops. The above feature accounts for the wide differences observed amongst crops in their responses when exposed to the same weather. The same holds good for photosynthesis also. Maize and Sorghum, Tobacco and Soybean are the most efficient, least efficient and average users of sunlight respectively. Similarly wheat is a more efficient user of CO_2 than rice while maize is the least efficient user of CO_2. While reduced solar radiation in the vegetative phase of rice has little influence on yield except under excessively cloudy conditions (Stansel. 1975), solar radiation from panicle initiation to maturity is linearly related to yield of rice (Islam and Morison, 1992).

4.10 Solar Radiation and Crop Yields

Monteith (1972, 1977) had shown that Net Primary Production is proportional to the Intercepted Photosynthetically Active Radiation, IPAR. The proportionality of total drymatter produced to amount of IPAR was first shown to be applicable to cereals (Gallagher and Biscoe, 1978) and later to a wide range of crops (Monteith, 1994). The term Radiation Use Efficiency, RUE is used to denote the slope of the relationship between the amount of drymatter produced and quantum of IPAR. Now, PAR is 0.45 times Global Solar Radiation, GSR (Monteith, 1965). So RUE can be expressed in terms of GSR also.

RUE has been shown to vary amongst crops (Sivakumar and Virmani, 1984; Goose et. al. 1986; Prince, 1991). RUE has a narrow range for crop ecosystems (Potter et. al. 1993) but shows a much wider range for natural ecosystems (Russel et. al, 1989). Within a crop species RUE can vary with varieties as in case of wheat (Mishra et. al.

2009), rice (Venkataraman, 2009) and soybean (Singh et. al, 2007). From eco-physiological considerations of production and respiratory use of photosynthates, as detailed above, unit area yield of a given cultivar can, a priori, be expected to be influenced by the temperature and radiation regimes experienced by it and its weather-dependant ontogeny.

The unit area economic yields of new cultivars have been superior to those they had replaced. The field-life duration at a place of the new high yielding, hybrid, cultivars of many crops like rice, wheat and maize are nearly the same as the old varieties. So explanation in terms of yield per day gets ruled out. The ratio of highest to lowest RUE of varieties is seen to be 1.05 in case of wheat (Mishra et. al. 2009). 1, 25 in case of rice (Venkataraman, 2009) and 1.22 in case of soybean (Singh et. al. 2007). Harvest Index (HI), which is the ratio of economic yield of the crop to the net drymatter produced, is genetically manipulable. Much of the increase in crop yields in the green revolution has been achieved through increase in HI of cultivars, which is around 60% in present- day grain cultivars (Hay, 1995). However, realising a HI exceeding 60% and quantum jumps in RUE through crop breeding are unlikely (Long, 2006). For many crop cultivars, the leaf-architecture of their canopy is well below the optimal for the theoretical maximum interception of 90% of solar radiation (Beadle and Long, 1985) and there is good scope for substantially increasing interception of solar radiation by cultivars through conventional breeding.

In light of the above, maximization of (i) interception of solar radiation by the crop canopy in its field life and (ii) efficiency of use of available light are required to bring about a quick increase in productivity of crops. In the above, both agronomic and genetic manipulations, as mentioned below would be required to be adopted.

Absorption of solar radiation can be deemed to be complete if the radiation received by the lowest leaf in a canopy is well above the light compensation point so as to ensure full opening of the stomata and adequate photosynthesis. Agrometeorological models relating to penetration of light and absorption of solar radiation in crop canopies of given height and leaf-architecture are available (Venkataraman and Krishnan, 1992). By determining the level of solar radiation required to be received by the lowest leaf-layer and considering the solar radiation likely to be available at the top of the crop canopy, dynamic agrometeorological models on penetration of solar radiation in a crop stand can be used to arrive at the morphology of crops, principally height, thickness and orientation of leaves and vertical distribution of total leaf area that would ensure total penetration of solar radiation for a given location and period with minimal possible duration of light saturated photosynthesis at each leaf-level. Crop Breeders can then use location-specific monthly solar radiation, either observed or computed, to evolve appropriate cultivars to maximise IPAR. The development and use of Bractless Cotton in ensuring greater penetration of light and wind into the canopy to minimize attack by pests is a case in point.

The agronomic strategy must ensure that the crops develop to reach the ground-shading stage in the quickest possible time with minimum possible number of leaves per plant to avoid mutual shading (Shlumukov et. al. 2001) which will lower photosynthetic rates of shaded leaves. For this, (a) adoption of proper (i) sowing dates (ii) orientation of direction of crop rows (iii) row spacing (iv) distance between plants in rows and (b) choice of cultivars would be necessary. In varietal selection, crop phenology plays an important part. For example, field-life durations of rice and groundnut are mainly determined by the duration of vegetative and reproductive phases respectively. Again, in mixed cropping the taller crop should be of the C4 type and the shorter one of the C3 type. To adopt both genetic and agronomic measures for improving IPAR, location-specific monthly data of solar radiation, either observed or derived, will be required.

4.11 Coping with Climate Change

At existing levels of Carbon Dioxide in the air, solar radiation on clear days in the Tropics is supra-optimal for photosynthesis (Grant et. al. 1999). In the envisaged scenario of climate change, the level of crop-usable CO_2 in the air is expected to increase by 50% by the middle of this century (Sinha, 1993). Model studies show that even with expected levels of reduction in solar radiation, increase in CO_2 levels should increase yields. Studies by Rajvel et. al. (2010) show that optimal level of CO_2 for net photosynthesis for maize and safflower will be twice the present level of Carbon Dioxide, Thus, in climate change, the expected reduced levels of solar radiation will still be super-optimal for photosynthesis.

Theoretically, enhanced Carbon Dioxide should lead to a decrease in mean daily global solar radiation (Zaitao et. al. 2004). Reduction in solar radiation at 2% per decade at many stations in India during the clear weather period of May to December has been reported (Shende and Chivate, 2000). Stanhill and Cohen (2001) report that during the past 50 years, in many industrial regions of the globe, solar radiation has decreased by 2.7% per decade.

In climate change studies, reduction in radiation of 1, 2 and 3 $MJ/m^2/day$ is assumed for the optimistic, normal and pessimistic future climatic scenario respectively. The above quantities of radiation would amount to various fractions of normal radiation in different areas and seasons. Studies in yield reduction based on decrease of fixed quantum of radiation will not be realistic. Studies based on percentage reduction in solar radiation from the normal would be more meaningful. Up to 20% reduction in solar radiation is seen to have little influence on crop yields (Stanhill and Cohen, 2001; Swain et. al. 2007). The reduction in radiation of 1, 2 and 3 $MJ/m^2/day$ would amount to 5, 10 and 15% of normal radiation respectively. So the envisaged reductions in reduction, due to an increase in anthropogenic aerosols and pollutants changing the optical properties of the atmosphere (Stanhill and Cohen, 2001) should not be of serious agricultural concern.

A distinction needs to be made between reduction in radiation under clear skies and decrease in radiation due to increased cloudiness. The concept of Critical Sunlight Period (Ronald and Cicerone, 2002) stipulates that reduction of bright sunshine hours during specified crop stages will reduce crop yields. For example reduction in sunlight in the 6 week period following panicle initiation in rice will reduce the grain number (Swain et. al. 2007), increase spikelet sterility (Sinha, 1993) and ill-filled spikelets (Vijayalakshmi et. al. 1991) and hence lower rice yields.

While models indicate that reduction in productivity will be proportional to solar radiation, experimental evidence suggests lesser sensitivity, especially under high radiation regimes in arid climate, due to variations in shade tolerance amongst crops (Stanhill and

Cohen, 2001). Reducing solar radiation of spring wheat by 50% by shading from anthesis to maturity reduced yield by only 10% (Fischer, 1975). Cowpea and bushbean are respectively the least and most tolerant to shading, Yield of soybean is not affected by shading up to 30% reduction in Sunlight (Eriksen and Whitney, 1984). Boreal aspen forests and temperate zone mixed deciduous forests can tolerate reduction upto 50% in solar radiation caused by increased cloudiness without any lowering of their capacity for carbon uptake (Lianhong et. al. 1999). Increase in cloudiness will increase night temperatures, which will lead to greater depletion of photosynthates on account of an increase in the maintenance respiration which varies amongst crops over a narrower range (Bishnoi, 1986).

Significant year to year variations at a place of solar radiation received during the maturity period of a given crop are quite common. In view of the above yields of important grain crops need to be studied in relation to the normal level of and percentage reduction of sunshine hours and solar radiation in various crop phases so as to assess the reduction in yield in terms of percentage of the normal yield. Such assessments will help in anticipating, on a crop-wise basis, the likely reductions in yield on account of decrease in solar radiation in future climate and help in devising measures, both agronomic and genetical, to mitigate the reductions in yields.

References

Allen, R.G.; Periera, L.S.; Raes, D. and Smith, M. 1998. Crop Evapotranspiration. Guidelines for computing crop water requirements. Irrigation and Drainage paper No. 56, Food and Agriculture Organisation (FAO), 300 pp.

Beadle, C.L. and Long, S.P. 1985. Photosynthesis- is it limiting to biomass production? Biomass. 8: 119-168.

Bishnoi. O.P. 1986. Solar radiation and productivity in India. I. Potential productivity. Mausam, 37: 501-506.

Blackman, F.F. 1905. Optima and limiting factors. Ann. Bot. (Old Series), 19: 281-296.

Eriksen, F.I. and Whitney, A.S. 1984. Effects of solar radiation regime and N_2 fixation of soybean, cowpea and bush bean. Agron, Jl. 76: 529-535.

Fischer R. A. 1975. Yield potential in dwarf spring wheat and the effect of shading. Crop Science, 15: 607-613.

Gallagher, J.N. and Biscoem P, V. 1978. Radiation absorption, growth and yield of cereals. Jl. of Agric. Sci. 91: 47-60.

Gangopadhyay, M.; Datar, S.V. and George, C.J. 1970. On the global solar radiation climate and evapotranspiration estimâtes of India. Ind. Jl. Meteorol. and Geophys. 21: 23-30.

Goose, G, et. al. 1986. Maximum dry matter production and solar radiation intercepted by a canopy. Agronomie. 6: 47-56.

Grant, R.F. et al. 1999. Crop water relations under different CO_2 and irrigation: Testing of ecosystems with the Free Air CO_2 Enrichment (FACE) experiment. Agric. and Forest Meteorol. 95: 27-51.

Hay, R.K.M. 1995. Harvest Index: A review of its use in plant breeding and crop physiology. Annals of Applied Biol. 126: 197-216.

Idso, S.B. and Jackson, R.D. 1969. Thermal radiation from the atmosphere. Jl. Geophys. Res. 74: 5397-5403.

Islam, M.S. and Morison, J.I.L. 1992. Influence of solar radiation and temperature on irrigated rice grain yield in Bangladesh. Field Crops Res. 30: 13-28.

Kale, R.S., 1951. Studies in infra-red radiation of the atmosphere. Part l-An instrument for the measurement of infra-red radiation during day and night. Jl. Sci. and Ind. Res.10B. 155-160.

Liannhong, G. et. al. 1999. Response of net ecosystem exchange of carbon dioxide to change in cloudiness: Results from two North American deciduous forests. Jl. of. Geophysical Res. 104: (No.D 24). 31421-31434 pp.

Linacre, E.T. 1968. Estimating the net-radiation flux. Agric. Meteorol. 5: 49-63.

Long, S.P.; Zhu, X.G.; Naidu, S.L. and Ort, D.R. 2006. Can improvement in photosynthesis increase crop yields? Plant, Cell and Environment. 29: 315-330.

Mani, A.; Chacko, O. and Iyer, N.V. 1965. Studies of terrestrial radiation fluxes at the ground in India. Ind. Jl. Meteorol. and Geophys. 16: 445-452.

Mishra, A.K.; Tripathi, P.; Pal, R.K. and Mishra, S.R. 2009. Light interception and radiation use efficiency of wheat varieties as influenced by number of irrigations. Jl. of Agrometeorol. 11: 140-143.

Monteith, J.L. 1965. Evaporation and Environment. Proc. XIXth Symposium of Society for Experimental Biology. Cambridge University Press. 205-234 pp.

Monteith, J. L. 1972. Solar radiation and productivity in tropical ecosystems. Journal of Applied Ecol. 9: 747–766.

Monteith, J. L. 1977. Climate and the efficiency of crop production in Britain. Philosophical transactions of the Royal Soc. Of London. Series B 281: 277-294.

Monteith, J.L. 1994. Validity of the connection between intercepted radiation and biomass. Agric. and Forest Meteorol. 68: 213-220.

Pearcy, R.W. and Bjorkman, O, 1993. Physiological Effects. In: (Ed, Lemon, R,). The response of plants to rising levels of atmospheric carbon dioxide. Westview, Boulder, Colorado, U.S.A. 65-106 pp.

Potter, C.S. et al. 1993. Terrestrial ecosystem production. A Process based on global, satellite and surface data. Global Biogeo-chemical Cycles. 7: 811-841.

Prince. S.D. 1991. A model of regional primary production for use with coarse resolution satellite data. International Jl. of Remote Sensing, 12: 1313-1330.

Rajvel, M. et al. 2010. Effect of diurnal variation of atmospheric and elevated levels of carbon-dioxide and photosynthetically active radiation on intercellular concentration and rate of photosynthesis in maize and safflower. Jl. of. Agrometeorol, 12: 1-7.

Ronald. L. S. and Cicerone, R.J. 2002. Photosynthetic allocation in rice plants: Food production or atmospheric methane? Proc. Nat. Acad. Sci. USA, 99: 11993-11995.

Russel, G.P.; Jarvis, P.G. and Moteith, J.L. 1989. Absorption of radiation by canopies and stand growth In; Russel, G.; Marshall, B. and Jarvis, P.G. (Eds). Plant Canopies: Their Growth, form and function. Cambridge Univ. Press, Cambridge, U.K. 21-39 pp.

Shende. R.R. and Chivate, V.R. 2000. Global and diffuse solar radiation exposures at Pune. Mausam 51: 349-358.

Shlummukov. L.R. et al. 2001. Establishment of far-red high irradiance responses in wheat through transgenic expression of an oat phytochrome a gene. Plant, Cell and Environment. 24: 703-712.

Singh, A.; Rao, V.U.M.; Singh, D, and Singh, R. 2007. Study on agrometeorological indices for soybean crop under different growing environments. Jl. of Agrometeorol. 9: 81-85.

Sinha, S.K. 1993. Response of tropical ecosystems to climate change. International Crop Sci. Pub. Crop Science Soc. America, Madison, U.S.A. 282-289 pp.

Sivakumar, M.V.K. and Virmani. S.M. 1984. Crop productivity in relation to interception of photsynthetically active radiation. Agric. and Forest Meteorol. 31: 131-141.

Smith, M., Allen, R, Monteith, J.L., Perriera, L., Periera, L.S. and Segeren, A. 1992. Report, Expert consultation on revision of FAO methodologies for crop water requirements, FAO. 60 pp.

Stanhill, S. and Cohen, S. 2001. Global dimming: A review of the evidence for a widespread and significant reduction in global radiation with discussion of its probable causes and possible agricultural consequences. Agricl. and Forest Meteorol. 107: 255-278.

Stansel, J.W. 1975. Effective utilization of sunlight. Pages 43-50 in Texas Agricultural Experimental Station, in cooperation with the U.S. Department of Agriculture. Six decades of rice research in Texas. Res. Monogr. 4.

Swain, D.K.; Herath, S.; Saha, S. and Dash, R.N. 2007. CERES-Rice model: Calibration. Evaluation and application for solar radiation stress assessment on rice production. Jl. of. Agromrteorol. 9: 138-148.

Venkataraman, S. 1977. Evaluation of mean daily net longwave radiation flux using hypergeometric data. Mausam, 28: 402-403.

Venkataraman, S. 2009. Some preliminary observations on varietal influences on yield components of rice. Jl. of Agrometeorol. 12: 206-207.

Venkataraman, S. and Krishnamurthy, V. 1972. Clear sky atmospheric radiation over India. International. Symp. Radiation Measurements Including Satellite Techniques, Sendai, Japan. 33-42.

Venkataraman, S. and Krishnan, A. 1992. Crops and Weather. Publication Indian Council of Agricultural Research, 586 pp.

Vijayalakshmi, C.; Radhakrishnan, C.; Nagarajan, M. and Rajendran, C. 1991. Effect of solar radiation deficit on rice crop productivity. Jl. Of Agron. and Crop Sci. 167: 184-187.

Zaitao. P; Segal. M; Raymond, W.A. and Eugene, S.T. On the potential change in solar radiation over the U.S. Due to increase of atmospheric greenhouse gases. Renewable Energy, 29: 1923-1928.

Table 4.3A Weekly sunlight hours northern hemispheric latitudes

Std. Week	0	5	10	15	20	25	30	35	40	45	50	55	60	65	70
1	12.0	11.7	11.5	11.1	10.9	10.5	10.2	9.8	9.3	8.7	8.1	7.3	6.1	3.9	0.0
2	12.0	11.7	11.5	11.2	10.9	10.6	10.3	9.9	9.5	8.9	8.3	7.4	6.3	4.3	0.0
3	12.0	11.7	11.5	11.2	10.9	10.7	10.4	10.0	9.6	9.0	8.5	7.7	6.6	4.8	0.7
4	12.0	11.8	11.6	11.3	11.0	10.8	10.6	10.2	9.8	9.3	8.8	8.1	7.2	5.7	1.0
5	12.0	11.8	11.6	11.3	11.1	10.9	10.7	10.3	10.0	9.5	9.1	8.5	7.6	6.3	3.8
6	12.0	11.9	11.7	11.4	11.2	11.0	10.9	10.5	10.3	9.9	9.5	9.1	8.2	7.1	4.9
7	12.0	11.9	11.7	11.5	11.3	11.1	11.0	10.7	10.5	10.2	9.8	9.4	8.8	8.0	6.6
8	12.0	11.9	11.9	11.6	11.5	11.3	11.2	10.9	10.8	10.5	10.3	9.9	9.5	8.8	7.7
9	12.0	11.9	11.8	11.7	11.6	11.5	11.4	11.2	11.1	10.9	10.7	10.5	10.1	9.6	8.8
10	12.0	11.9	11.9	11.8	11.7	11.7	11.6	11.5	11.4	11.3	11.1	11.0	10.7	10.4	9.9
11	12.0	12.0	11.9	11.9	11.9	11.8	11.8	11.7	11.7	11.6	11.6	11.5	11.4	11.2	11.0
12	12.0	12.1	12.0	12.0	12.1	12.0	12.1	12.0	12.1	12.1	12.1	12.1	12.1	12.1	12.1
13	12.0	12.1	12.1	12.1	12.1	12.1	12.2	12.2	12.3	12.3	12.3	12.5	12.5	12.6	12.8
14	12.0	12.1	12.1	12.1	12.3	12.3	12.4	12.5	12.6	12.7	12.7	13.0	13.1	13.5	13.9
15	12.0	12.1	12.2	12.3	12.4	12.5	12.6	12.8	13.0	13.1	13.3	13.7	13.9	14.4	13.1

Table 4.3A *Contd....*

Std. Week	0	5	10	15	20	25	30	35	40	45	50	55	60	65	70
16	12.0	12.1	12.2	12.3	12.5	12.7	12.8	13.0	13.3	13.5	13.7	14.1	14.6	15.3	16.3
17	12.0	12.1	12.3	12.4	12.6	12.8	13.0	13.2	13.5	13.8	14.0	14.6	15.1	15.9	17.5
18	12.0	12.2	12.3	12.5	12.7	13.0	13.2	13.5	13.8	14.1	14.4	15.0	15.7	16.8	18.9
19	12.0	12.3	12.4	12.6	12.8	13.1	13.3	13.7	14.0	14.4	14.8	15.5	16.3	17.6	20.2
20	12.0	12.3	12.5	12.7	12.9	13.3	13.5	13.9	14.3	14.8	15.3	16.0	17.1	18.5	21.5
21	12.0	12.3	12.5	12.7	13.0	13.3	13.6	14.0	14.4	14.9	15.5	16.0	16.9	19.1	22.1
22	12.0	12.3	12.5	12.8	13.1	13.4	13.7	14.1	14.5	15.1	15.7	16.5	17.6	19.7	22.7
23	12.0	12.3	12.5	12.9	13.1	13.5	13.8	14.2	14.7	15.3	15.9	16.8	18.0	20.4	22.3
24	12.0	12.3	12.6	12.9	13.2	13.5	13.9	14.3	14.8	15.4	16.1	17.1	18.4	21.1	24.0
25	12.0	12.3	12.5	12.9	13.1	13.5	13.9	14.3	14.7	15.3	16.0	17.0	18.3	20.7	24.0
26	12.0	12.3	12.5	12.9	13.1	13.5	13.9	13.2	14.7	15.3	15.9	16.9	18.1	20.4	24.0

Table 4.3A *Contd....*

Std. Week	0	5	10	15	20	25	30	35	40	45	50	55	60	65	70
27	12.0	12.3	12.5	12.9	13.1	13.5	13.9	14.2	14.7	15.3	15.9	16.8	18.0	20.2	24.0
28	12.0	12.3	12.5	12.9	13.1	13.4	13.8	14.1	14.6	15.1	15.7	16.5	17.6	19.5	24.0
29	12.0	12.3	12.5	12.9	13.1	13.3	13.7	14.0	14.5	14.9	15.5	16.4	17.4	19.2	23.2
30	12.0	12.2	12.4	12.8	13.0	13.2	13.6	13.9	14.3	15.0	15.2	15.9	16.8	18.5	21.5
31	12.0	12.2	12.4	12.7	12.9	13.1	13.5	13.7	14.1	14.5	15.0	15.7	16.5	17.9	20.8
32	12.0	12.1	12.3	12.6	12.8	13.0	13.3	13.5	13.5	14.2	14.7	15.2	15.9	17.1	19.2
33	12.0	12.1	12.3	12.5	12.7	12.9	13.1	13.3	13.6	13.9	14.3	14.7	15.3	16.2	17.6
34	12.0	12.1	12.3	12.4	12.5	12.7	12.9	13.0	13.3	13.5	13.8	14.1	14.6	15.3	16.4
35	12.0	12.1	12.2	12.3	12.4	12.5	12.7	12.7	12.9	13.1	13.3	13.5	13.9	14.5	15.2
36	12.0	12.1	12.1	12.2	12.3	12.3	12.5	12.5	12.5	12.7	12.8	12.9	13.2	13.6	14.0
37	12.0	12.0	12.1	12.1	12.1	12.2	12.2	12.3	12.3	12.4	12.4	12.5	12.7	12.9	13.1
38	12.0	12.0	12.0	12.0	12.0	12.0	12.0	12.0	12.0	12.0	12.0	12.0	12.0	12.0	12.0
39	12.0	11.9	12.0	12.0	11.9	11.8	11.9	11.9	11.7	11.8	11.6	11.8	11.6	11.5	11.3
40	12.0	11.9	11.9	11.9	11.7	11.6	11.6	11.7	11.4	11.3	11.1	10.9	10.8	10.7	10.4

Table 4.3A *Contd...*

Std. Week	0	5	10	15	20	25	30	35	40	45	50	55	60	65	70
41	12.0	11.9	11.8	11.7	11.6	11.8	11.4	11.3	11.1	10.9	10.6	10.4	10.2	9.5	8.9
42	12.0	11.9	11.8	11.7	11.5	11.3	11.2	11.0	10.8	10.6	10.2	9.9	9.4	8.7	7.7
43	12.0	11.8	11.7	11.6	11.4	11.2	11.1	10.8	10.5	10.3	9.9	9.5	8.9	8.0	6.5
44	12.0	11.8	11.7	11.5	11.3	11.1	10.9	10.6	10.3	10.0	9.5	9.1	8.4	7.9	5.3
45	12.0	11.7	11.6	11.4	11.1	10,9	10.7	10.3	10.0	9.7	9.1	8.6	7.7	9.5	3.8
46	12.0	11.7	11.5	11.3	11.0	10.7	10.5	10.1	9.7	9.3	8.7	8.1	7.1	5.6	2.3
47	12.0	11.7	11.5	11.3	10.9	10.7	10.4	10.0	9.6	9.1	8.5	7.8	6.7	4.9	1.7
48	12.0	11.7	11.5	11.2	10.9	10.6	10.3	9.9	9.5	9.0	8.4	7.6	6.5	4.5	1.3
49	12.0	11.7	11.5	11.2	10.9	10.6	10.2	9.8	9.4	8.8	8.2	7.3	6.1	3.7	0.4
50	12.0	11.7	11.4	11.1	10.8	10,5	10.1	9.7	9.5	8.6	8.0	7.0	5.7	3.0	0.2
51	12.0	11.7	11.5	11.1	10.9	10.5	10.1	9.7	9.3	8.7	8.0	7.2	5.7	3.1	0.1
52	12.0	11.7	11.5	11.1	10.9	10.5	10.1	9.7	9.3	8.7	8.0	7.1	5.8	3.3	0.0

Table 4.3B Weekly daylight hours southern hemispheric latitudes

Std. Week	0	5	10	15	20	25	30	35	40	45	50	55	60	65	70
1	12.0	12.2	12.5	12.9	3.1	13.5	13.7	14.2	14.7	15.3	15.9	16.7	'7.0	20.1	24.0
2	12.0	12.3	12.5	12.8	13.1	13.4	13.7	14.1	14.5	15.1	15.7	16.6	7.7	19.7	24.0
3	12.0	12.3	12.5	12.8	13.1	13.3	13.6	14.0	14.4	15.0	15.5	16.3	17.4	19.1	23.4
4	12.0	12.3	12.5	12.7	13.0	13.2	13.5	13.9	14.2	14.7	15.2	15.9	16.9	18.4	22.1
5	12.0	12.2	12.4	12.7	12.9	13.1	13.3	13.7	14.0	14.5	14.9	15.6	16.3	17.7	20.7
6	12.0	12.1	12.3	12.6	12.8	13.0	13.1	13.5	13.7	14.1	14.5	15.1	15.8	16.9	19.1
7	12.0	12.1	12.3	12.5	12.7	12.9	13.0	13.3	13.5	13.8	14.2	14.6	15.2	16.0	17.4
8	12.0	12.1	12.3	12.4	12.5	12.7	12.8	13.1	13.2	13.5	13.7	14.1	14.5	15.2	16.3
9	12.0	12.1	12.2	12.3	12.4	12.5	12.6	12.8	12.9	13.1	13.3	13.5	13.9	14.4	15.2
10	12.0	12.1	12.1	12.2	12.3	12.3	12.4	12.5	12.6	12.7	12.9	13.0	13.3	13.6	14.1
11	12.0	12.1	12.1	12.1	12.1	12.2	12.2	12.3	12.3	12.4	12.4	12.5	12.6	12.8	13.0
12	12.0	12.0	12.0	12.0	11.9	12.0	11.9	12.0	11.9	11.9	11.9	11.9	11.9	11.9	11.9
13	12.0	11.9	11.9	11.9	11.9	11.9	11.8	11.8	11.7	11.7	11.7	11.5	11.5	11.4	11.2
14	12.0	11.9	11.9	11.8	11.7	11.7	11.5	11.5	11.4	11.2	11.2	10.9	10.7	10.4	9.9
15	12.0	11.9	11.7	11.7	11.5	11.5	11.4	11.2	11.0	10.8	10.6	10.3	10.0	9.5	8.7

Table 4.3B Contd....

Std. Week	0	5	10	15	20	25	30	35	40	45	50	55	60	65	70
16	12.0	11.9	11.9	11.7	11.5	11.3	11.2	11.0	10.8	10.5	10.3	9.8	9.5	8.8	7.7
17	12.0	11.8	11.7	11.6	11.4	11.2	11.0	10.8	10.5	10.2	10.0	9.4	8.9	8.1	6.5
18	12.0	11.8	11.7	11.5	11.3	11.0	10.8	10.5	10.2	9.9	9.6	9.0	8.3	7.2	5.1
19	12.0	11.7	11.6	11.4	11.2	10.8	10.6	10.3	10.0	9.5	9.1	8.4	7.4	6.7	3.6
20	12.0	11.7	11.5	11.3	11.1	10.7	10.5	10.1	9.8	9.2	8.7	8.0	7.1	5.5	2.6
21	12.0	11.7	11.5	11.3	11.0	10.7	10.4	10.0	9.7	9.5	8.5	7.7	6.9	4.9	1.9
22	12.0	11.7	11.5	11.2	10.9	10.6	10.3	9.9	9.5	9.9	8.3	7.5	6.4	4.3	1.3
23	12.0	11.7	11.5	11.1	10.9	10.5	10.2	9.8	9.3	8.7	8.1	7.2	6.0	3.6	0.7
24	12.0	11.7	11.4	11.1	10.8	10.5	10.1	9.7	9.2	8.6	7.9	6.9	5.6	2.9	0.0
25	12.0	11.7	11.4	11.1	10.8	10.5	10.1	9.7	9.2	8.6	8.0	7.0	5.7	3.2	0.0
26	12.0	11.8	11.5	11.1	10.9	10.5	10.1	9.8	9.3	8.7	8.1	7.1	5.8	3.5	0.0

Table 4.3B Contd....

83

Std. Week	0	5	10	15	20	25	30	35	40	45	50	55	60	65	70
27	12.0	11.8	11.5	11.1	10.9	10.5	10.1	9.8	9.3	8.7	8.1	7.2	6.0	3.8	0.0
28	12.0	11.9	11.5	11.1	10.9	10.6	10.2	9.8	9.4	8.8	8.3	7.3	6.2	4.1	0.0
29	12.0	11.9	11.5	11.1	10.9	10.7	10.3	10.0	9.5	9.0	8.5	7.6	6.6	4.7	0.8
30	12.0	11.9	11.6	11.2	11.0	10.8	10.4	10.3	9.7	9.3	8.7	7.9	7.0	5.6	1.3
31	12.0	11.9	11.6	11.3	11.1	10.9	10.5	10.3	9.9	9.5	9.0	8.3	7.5	6.1	3.2
32	12.0	11.9	11.7	11.4	11.2	11.0	10.7	10.5	10.1	9.8	9.3	8.8	8.1	6.9	4.8
33	12.0	11.9	11.7	11.5	11.3	11.1	10.9	10.7	10.4	10.1	9.7	9.3	6.7	7.8	6.4
34	12.0	11.9	11.7	11.6	11.5	11.3	11.1	11.0	10.7	10.5	10.2	9.9	9.4	8.7	7.6
35	12.0	11.9	11.8	11.7	11.6	11.5	11.3	11.3	11.1	10.9	10.7	10.5	10.1	9.6	8.8
36	12.0	11.9	11.9	11.8	11.7	11.6	11.5	11.5	11.3	11.2	11.0	10.9	10.6	10.1	9.6
37	12.0	12.0	11.9	11.9	11.9	11.8	11.7	11.7	11.7	11.6	11.6	11.4	11.3	11.0	10.9
38	12.0	12.0	12.0	12.0	12.0	12.0	12.0	12.0	12.0	12.0	12.0	12.0	12.0	11.9	12.0
39	12.0	12.1	12.0	12.0	12.1	12.2	12.1	12.1	12.3	12.2	12.4	12.3	12.4	12.5	12.7
40	12.0	12.1	12.1	12.2	12.3	12.4	12.4	12.4	12.6	12.7	12.9	12.9	13.2	13.5	14.0
41	12.0	12.1	12.2	12.3	12.4	12.5	12.6	12.7	12.9	13.1	13.4	13.6	13.9	14.5	15.5

Table 4.3B Contd....

Std. Week	0	5	10	15	20	25	30	35	40	45	50	55	60	65	70
42	12.0	12.1	12.2	12.3	12.5	12.7	12.8	13.1	13.2	13.4	13.8	14.1	14.6	15.3	16.3
43	12.0	12.1	12.3	12.4	12.7	12.7	12.9	13.1	13.4	13.7	14.1	14.4	14.9	15.8	17.5
44	12.0	12.2	12.3	12.5	12.7	12.9	13.1	13.4	13.7	14.0	14.5	14.9	15.6	16.7	18.9
45	12.0	12.3	12.4	12.6	12.9	13.1	13.3	13.7	14.0	14.3	14.9	15.4	16.3	17.5	20.3
46	12.0	12.3	12.5	12.7	13.0	13.3	13.5	13.9	14.3	14.7	15.3	15.9	16.9	18.4	21.7
47	12.0	12.3	12.5	12.7	13.1	13.3	13.6	14.0	14.4	14.9	15.5	16.2	17.7	19.8	22.3
48	12.0	12.3	12.5	12.8	13.1	13.4	13.7	14.1	14.5	15.0	15.6	16.4	17.6	19.6	22.7
49	12.0	12.3	12.5	12.9	13.1	13.5	13.8	14.2	14.7	15.2	15.8	16.8	17.9	20.3	23.3
50	12.0	12.3	12.6	12.9	13.2	13.5	13.9	14.3	14.8	15.4	16.0	17.0	18.3	21.0	23.8
51	12.0	12.3	12.5	12.9	13.1	13.5	13.9	14.3	14.7	15.3	16.0	16.9	18.2	20.8	23.9
52	12.0	12.3	12.5	12.9	13.1	13.5	13.8	14.2	14.7	15.3	16.0	16.9	18.1	20.6	24.0

Table 4.4 Saturation vapour pressure, V.P. over water, in mm for different temperatures, TC

TC	VP	TC	VP	TC	VP	TC	VP	TC	VP
1.0	4.93	10.5	9.53	20.0	17.53	29.5	30.92	39.0	52.43
1.5	5.11	11.0	9.85	20.5	18.09	30.0	31.82	39.5	53.86
2.0	5.29	11.5	10.18	21.0	18.65	30.5	32.75	40.0	55.32
2.5	5.48	12.0	10.52	21.5	19.23	31.0	33.70	40.5	56.81
3.0	5.69	12.5	10.87	22.0	19.83	31.5	34.67	41.0	58.33
3.5	5.89	13.0	11.23	22.5	20.45	32.0	35.66	41.5	59.89
4.0	6.10	13.5	11.60	23.0	21.07	32.5	36.68	42.0	61.49
4.5	6.31	14.0	11.99	23.5	21.72	33.0	37.73	42.5	63.13
5.0	6.54	14.5	12.38	24.0	22.38	33.5	38.80	43.0	64.80
5.5	6.77	15.0	12.79	24.5	23.06	34.0	39.89	43.5	66.50
6.0	7.01	15.5	13.21	25.0	23.76	34.5	41.02	44.0	68.23
6.5	7.26	16.0	13.63	25.5	24.47	35.0	42.17	44.5	70.04
7.0	7.51	16.5	14.08	26.0	25.21	35.5	43.35	45.0	71.87
7.5	7.78	17.0	14.53	26.5	25.97	36.0	44.56	45.5	73.74
8.0	8.05	17.5	15.00	27.0	26.74	36.5	45.79	46.0	75.65
8.5	8.33	18.0	15.48	27.5	27.53	37.0	47.06	46.5	77.60
9.0	8.61	18.5	15.97	28.0	28.35	37.5	48.36	47.0	79.60
9.5	8.90	19.0	16.48	28.5	29.18	38.0	49.69	47.5	81.64
10.0	9.21	19.5	17.00	29.0	30.05	38.5	51.05	48.0	83.72

Table 4.5 Thermal back radiation ($s\, T_k^4$) - Calories/cm^2/day at various temperatures T_c. $S = 1.1711 \times 10^{-7}$ and $T_k = T_c + 273.16$.

TC	$s\, T_k^4$	TC	$s\, T_k^4$	TC	$s\, T_k^4$	TC	$s\, T_k^4$	TC	$s\, T_k^4$
1.0	662	10.5	759	20.0	865	29.5	983	39.0	1113
1.5	667	11.0	764	20.5	871	30.0	990	39.5	1120
2.0	672	11.5	769	21.0	877	30.5	996	40.0	1127
2.5	677	12.0	775	21.5	883	31.0	1003	40.5	1134
3.0	682	12.5	780	22.0	889	31.5	1009	41.0	1141
3.5	686	13.0	786	22.5	895	32.0	1016	41.5	1149
4.0	691	13.5	791	23.0	901	32.5	1023	42.0	1156
4.5	696	14.0	796	23.5	908	33.0	1030	42.5	1163
5.0	701	14.5	802	24.0	914	33.5	1036	43.0	1171
5.5	707	15.0	808	24.5	920	34.0	1043	43.5	1178
6.0	712	15.5	813	25.0	926	34.5	1050	44.0	1186
6.5	717	16.0	819	25.5	932	35.0	1057	44.5	1193
7.0	722	16.5	825	26.0	939	35.5	1063	45.0	1201
7.5	727	17.0	830	26.5	945	36.0	1071	45.5	1208
8.0	732	17.5	836	27.0	951	36.5	1077	46.0	1216
8.5	737	18.0	842	27.5	957	37.0	1084	46.5	1223
9.0	743	18.5	848	28.0	964	37.5	1091	47.0	1231
9.5	748	19.0	854	28.5	970	38.0	1098	47.5	1239
10.0	753	19.5	860	29.0	977	38.5	1106	48.0	1247

CHAPTER 5

Agrometeorology of Crop Water Usage

Frequently references are made to higher unit area yields of irrigated crops in high latitudinal countries compared to tropical countries. This gives the impression that something is radically wrong with the agricultural planning in the tropics. Across locations, yields of crops vary in a season. However, when such yields are expressed as kg/ha/day of field crop-life, the unit area yields per day are seen to be the same (Swaminathan, 1968; Muchinda and Venkataraman 1988). Thus, higher yields in the temperate zones are due to longer field-occupancy by crops. The picture that emerges from the above is that, in a given season for a given crop variety, the crop has to occupy the land for a longer time for higher yields. But supra-optimal solar radiation for photosynthesis and high temperatures (Grant et. al. 1998; Ritchie and NeSmith, 1991) reduce both daily biomass production and field-life of crops in the tropics compared to temperate zone areas. By corollary it means that in a given time scale more time is available for raising crops in the low latitudes subject to availability of water as temperature hardly restrains raising of crops in the tropics. Thus, the assured way to increase crop production in low latitudes is through increase in net irrigated acreage and maximisation of gross irrigated acreage by relay cropping. So, in the tropics, embarking on projects for augmentation of quantum of irrigation water supplies by construction of dams across rivers was a sound and rational strategy to ensure food security.

However, charging farmers for irrigation on the basis of area irrigated and not quantum of water supplied have led to farmers over-irrigating their crops and growing of water guzzling crops in unsuitable soils and in areas and periods of high water duty. Thus, the irrigation projects have not realised the envisaged potential in gross irrigated acreage and virgin soils opened up for irrigation have become saline in many a developing country. Lives and capacities of constructed dams have steadily declined due to excessive sedimentation arising from deforestation in their catchment areas.

A very high fraction of groundwater is at present used in raising commercial crops in high water duty areas and seasons. In many cases groundwater is being used as a sole source of irrigation to increase net irrigated acreage. The above are the most potent factors in alarming drops in groundwater levels. A radioactive dating of groundwater may show them to be many centuries old. Thus, a resource accumulated over many centuries is being depleted in a matter of decades. In India the point of over-exploitation of groundwater to the point of continuously decreasing levels appear to have been reached in 1997 itself (Garg and Hassan, 2007). The mooted idea of social ownership of groundwater reserves is not possible. Groundwater recharge is highly erratic in time and space and its augmentation by artificial recharge cannot be specifically planned.

The traditional dryland crop varieties had the ability to curtail transpiration under inadequate soil moisture (Venkataraman, 1981a) and had some fodder value even when they failed as grain crops because of their tallness. However, the new dwarf cultivars, that played a major role in the green revolution, had no physiological senescence (Venkataraman, 1985). Therefore, their peak water consumption covered the maturity period during which dryland crops mostly suffer from soil moisture stress. The new hybrid cultivars were prima facie, not suitable for use under rainfed conditions. The temporal distribution of rainfall is outside human control. However, collection of surface runoff from rains in percolation tanks and use of the harvested rainwater for drought-alleviation of crops during rainless periods have been successful at many locations. But for want of an infrastructure for rainfall harvesting, many savable dryland crops have been and are being lost.

5.1   Need for Optimal and Conjunctive use of Water Resources

In the context of increasing the rate of crop production in a sustained manner, the above mentioned misuse of surface irrigation, squandering of groundwater and non-maximal use of rainfall present an ominous picture and calls for a proper planning and execution of a programme of optimal and conjunctive use of our water resources as spelt out below.

More land cannot be opened up for agriculture. In light of the drop in the rate of increase in yields of irrigated crops in the erstwhile areas of green revolution in the last 15 years in India (Gadgil and Rao, 2000) the task of increasing unit area productivity of crops will be a daunting task. Irrigation projects to increase net irrigated acreage to the required level will be time consuming and costly. There is justifiable skepticism for creating, at great cost, additional water resources for crop production. Projects like interlinking of national rivers will pose problem of determining and sharing surpluses and deficits in river basins on an operational and yearly basis and will not solve water shortage problems of elevated plateau regions. Under the circumstances, increasing the gross area under irrigation through optimal use of surface irrigation water emerges as the main method for achieving agricultural sustainability in the short term.

The need for conjunctive use of water resources stems from the following. Long duration irrigated crops pass through the rainy season and effective use of rainfall by agronomic techniques can considerably reduce irrigation water needs for the crop and limit application of irrigation to critical periods to avoid water stress. Even in case of clear season irrigated crops, local use of groundwater for maturing a relay crop, which is likely to encounter warm weather, may be necessary to realise potential yields. Again, irrigation water is not available for establishment of crops in the pre-rainy period and waiting for canal irrigation will mean valuable loss of time and crop yield. In such cases establishment of community nurseries or use of drip-irrigation using ground water will be a necessity. In case of dryland crops, in situ rainfall often cannot meet crop water needs and use of rainfall-harvested storages and/or ground water will be necessary. The early tea crop in Assam needs pre-monsoon rains (Venkataraman and Krishnan, 1992) and in their absence sprinkler irrigation fed from groundwater reserves can help maintain yield levels of the crop. Sprinkler irrigation is best done at night to minimise evaporative loss of spray and given only for a ground covering crop. Sprinkler irrigation is a crop and not a water saving measure. Even Dew can contribute to water needs of Rabi crops in mild-winter areas (Raman et. al. 1973).

There are indications that our utilisable water resources might be significantly overestimated. The ongoing specter of global warming will lead to melting of glaciers. Initially flow in rivers fed by glaciers will increase and the associated increase in thickness of fresh water over sea surfaces will lead to an increase in rainfall. Gradually both river-flow and rainfall will decrease leading to water scarcity. Global warming will lead to (i) increase in level of carbon dioxide concentration which will reduce crop transpiration and (ii) reduction in evaporative power of air, which will further reduce crop water needs. The indications are that the percentage reduction in crop water need will be much more than the percentage reduction in field life of all crops in all areas and seasons. Thus, savings in irrigation water can, in fact, be used to increase irrigated acreage and gross output of the crop with existing irrigation potential even under global warming.

5.1.1   Role of Agroclimatology

Water requirements of crops, supplementary irrigation needs in the rainy season, erosion of top soil, groundwater recharge and temporal march of quantum and times of surface runoff and root zone moisture accretion from rains are weather-controlled. Thus, climatic data can be used for location-specific planning of (i) optimum irrigation scheduling of irrigated relay crops (ii) moisture availability periods for dryland crop culture and (iii) of measures for (a) rainfall harvesting (b) groundwater augmentation and (c) control of soil erosion.

5.2   Evapotranspiration

The transpirstory need of the crop, T constitutes the minimum water need of the crop to be met with. When moisture provisioning is done through the soil surface, evaporation E is inevitable. The combined loss of T and E from cropped fields is Evapotranspiration ET, which is synonymous with the term Consumptive Use employed by irrigation engineers. Though E and T occur simultaneously they are differently affected by the ambient air conditions as enumerated later.

5.2.1   Estimation

The aerodynamic profile method, the energy balance approach and the combination method which is an integration of the energy balance and aerodynamic methods are the three mainly used for estimating evapotranspiration of crops.

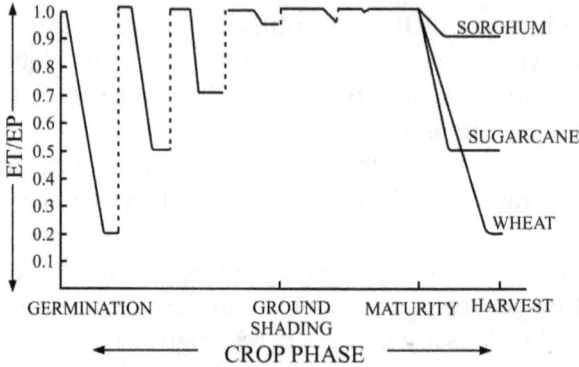

Fig. 5.1 March of clear weather relative evapotranspiration

5.2.1.1 Aerodynamic Profile Method

This technique is concerned with the transfer of water vapour between two levels in the air at a small distance above the crop surface. It uses the basic flux-gradient relationship of any atmospheric property, whose average concentration varies with height and gives rise to its transport. It involves measurement of specific humidity and wind speed at 2 heights with the proviso that the heights of measurements be less than 1/100 times the downwind fetch of the crop and the heights for specific humidity and wind speed be the same. This method is based on the assumption that wind speed drops to zero at the soil surface of the crop. However, this assumption is valid only for neutral conditions of stability of the atmosphere. Also, for aerodynamically rough crops the surface of zero wind, called zero Plane Displacement, d is pushed upwards. For short statured vegetation d can be taken as fixed fraction of crop height, called roughness length, z_o. For tall and dense vegetation d is pushed near the top of the crop. Again, observations of at least 5 heights separated by a meter each is required to determine d. To overcome the above problems determination of a Drag Coefficient, C_s during neutral stability conditions and use of the values of C_s so determined for other hours, has been suggested (Deacon and Swinbank, 1956). This involves measurement of wind speed U_s at a height Z_s as close as practicable to the crop surface. Thus, in the Aerodynamic Profile method, the equation used to determine ET is:

$$ET = C_s U_s^2 (q_1-q_2)/(u_2-u_1) \text{ and } C_s = K^2 \{(u_2/u_1)-1)^2\}/ (\log Z_2/Z_1)^2$$

In the above, q_1 and q_2 are specific humidities recorded at heights Z_1 and Z_2, u_1, u_2 and us are wind speeds recorded at heights Z_1, Z_2, and Z_s respectively and K is the Von Karman's constant equal to 0.41. In the above Z_2 may be high enough as dictated by the crop fetch downwind.

Drawback

The method is suited only to pastures and crops of short or medium height and not for long flexible crops which bend over to varying degrees with changes in wind speed. Observations of wind speeds at five heights separated by a meter each are required to determine the zero plane displacement. The crop fetch required to carry out the wind speed observations are not available at a research station. The method requires highly sophisticated equipment and computation of hourly values for integration into daily determinations of ET.

5.2.1.2 Energy Balance Approach

This method is based on the principle of conservation of energy. The energy available for ET is found by deducting the energy stored in the crop canopy and conducted into and out of the soil from the net radiation available at the crop surface and assuming (Bowen, 1926) that the resultant energy is shared between ET and conduction in the same ratio as the gradient of vapour pressure to temperature existing close to and above the crop canopy. The above is known as the Bowen Ratio, B

The Energy Budget equation can, therefore be written as follows:

$$ET = \frac{Rs(1-r) + 0.97\left(AR - ST_s^4\right) - G - Q}{1+B} \quad \text{and} \quad B = \frac{(T_1 - T_2) * (0.61) * P}{(e_1 - e_2) * 1000}$$

In the above Rs is solar radiation, r is the albedo, AR is the atmospheric long wave radiation, S is the Stefan-Boltzman constant, Ts is the mean air temperature in degrees absolute, G is energy storage in the canopy, Q is the heat flux into or out of soil, T_1 and T_2 and e_1 and e_2 are temperatures and vapour pressure measured in degrees centigrade and in millibars respectively at heights Z_1 and Z_2 and P is the station level pressure in millibars.

5.2.1.3 The Combination Method

This approach, postulated by Penman (1948), combines the energy budget and aerodynamic methods. In this ET is given by the relation:

$$ET = (Q_n \Delta/\gamma + E_a/\Delta \, (\gamma + 1) \text{ and } E_a = f(u) * (e_a - e_d)$$

In the above Q_n is net radiation, Δ/γ is the dimensionless ratio of the slope of the saturated vapour pressure versus temperature curve, u is wind speed at 2 meters height, e_a is saturation vapour pressure corresponding to mean air temperature and e_d is vapour pressure of air.

Drawback

Mainly suited for use with short stature and aerodynamically smooth crops and not with agricultural crops that are tall and/or aerodynamically rough.

5.2.1.4 The Resistance Approach

In this the various resistances encountered by the water vapour in moving from a level where wind velocity is zero to reach the atmosphere are taken into account. These are: (i) the external or aerodynamic resistance (sec/cm) indicative of the time taken by unit volume of water to exchange moisture with unit area of leaf surface (ii) internal resistance (sec/cm) provided by stomatal pores, cuticle and cell wall of leaves to diffusion of water within the leaves and (iii) atmospheric diffusive resistance.

In this method Potential Transpiration, PT, is given by the following relationship, namely:

$$\frac{PT}{Q_n} = \frac{\left[(\Delta/\gamma) + (r_i/r_a)\right]}{\left[(\Delta/\gamma + 1 + r_s/r_a)\right]}$$

In the above (i) Q_n is net radiation, $(Cal/cm^2/day)$, (ii) Δ/γ is the dimensionless ratio as mentioned above, (iii) r_i is the atmospheric diffusive resistance, (iv) r_a is the external or aerodynamic resistance, and (v) r_s is the internal resistance. The unit for above resistances is seconds per Centimeter. In the above r_i is given by the relation $r_i = PC \, (e_a - e_d)/\times Q_n$. γ where (i) P is the density of air above the crop surface with a value of $1.19 \times 10^{-3} \, gm/cm^3$, (ii) C is specific heat of air at constant pressure = 0.24 calories/degrees centigrade per gram, (iii) $(e_a - e_d)$ is vapour pressure deficit in millibars, (iv) Q_n is $cal/cm^2/sec$ and $\gamma = 0.66 \, mb/°C$.

Values of 1.1 and 0.36 for the external or aerodynamic resistance, r_a, of short and tall crops respectively and a value of 0.5 for r_s for both short and tall crops have been suggested by Monteith (1965). When crop is transpiring at the potential rate, evaporation form the soil surface, will be, as detailed above, 25% of PT. So PET can be taken as 1.25 times PT.

5.3   Field Measurements of Water Requirements of Crops

The approaches outlined for estimation of ET call for sophisticated instrumentation and very high technical skills and hence can be at a few locations and for brief periods and at the end of all that labour remain only estimates which need to be verified by ground-truth. In fact some significant results emerging from the above approaches is due to availability of corroborative field data on ET. For real-time applications, the need is for factually determined daily ET data on a crop-wise and region-wide basis.

5.3.1   Irrigation Trials

The term "consumptive use" used by irrigation engineers is synonymous with evapotranspiration. To determine consumptive use, irrigation engineers conduct, for prospective crops in the command area of irrigation projects, field trials to assess the right interval of irrigation and the amounts to be given at each irrigation so that there is no loss of water by percolation beyond the root zone of crops. In such trials, randomised replication of treatments consisting of a large number of combinations of quantum of irrigation water applied and intervals of irrigation are used. The combination that gives the least water usage and best yield is taken as the right one for irrigation scheduling.

Drawbacks

The treatments are chosen in a "Hit or Miss" approach, which necessitates use of small sized treatment plots. Maintenance of water-tightness of small plots is over-taxing and costly. Because of variations in interval of irrigation, the small plots will be subject, in a disorderly and random manner, to the "Oasis" and "Island" effects. This vitiates the water needs of crops in small sized plots. Such field trials can be conducted only after the construction of the dams. This means some gestation period for availability of results for field applications. Irrigation scheduling for prospective crops in virgin areas waiting to be opened up for irrigation can at best be gauged from recommendations for same crops in climatically homologous area.

5.3.2   Soil Moisture Profiling

5.3.2.1   Manual Methods

5.3.2.1.1 Soil Augers

In this method soil samples are collected by soil augers at a number of depths from surface to a little beyond the deepest roots after irrigation when percolation has ceased and just before the next irrigation is due. The soil samples are oven dried at 105 °C to get grams of water per 100 grams of dry soil. To convert the gravimetric moisture data to amount of moisture contained in the column, additional measures of profiles of bulk density in the soil are required. In this method there is an inevitable delay in obtaining the soil moisture content data and there are also the uncertainties involved due change of place for each measurement.

5.3.2.1.2 Core Sampling

In this method undisturbed cores of samples from surface to a little beyond the deepest roots are taken by manual or powered samplers lined with removable containers. The method avoids the need for measurement of bulk density profiles. However, the same drawbacks of the soil auger method persist.

5.3.2.1.3 Sensors

In this blocks of gypsum, nylon and fiber glass, either singly or in combination, with electrodes spaced in them at specified separation are used. The resistance between the electrodes vary with the moisture content of the soil. However, the calibration between soil moisture and the resistance between electrodes shows hysteresis. The blocks can be used only over a limited moisture range. Temperature probes in which the time taken to register a given rise of temperature is sought to be related to moisture content of the soil. This method is not suitable for saline soils and calibration varies with soil packing.

Pressure gauges, in which the suction set up by movement of water from or to a porous cup buried in the soil is measure of soil moisture content of soil, are also used. However, this method is suited only for use in the higher ranges of soil moisture.

Drawback

The quantum of Evapotranspiration (ET), from the entire soil profile is too small to be reflected in the soil moisture values on a daily basis. The heavier the soil the greater is the lag in the reflection of ET loss in soil moisture measurements. So method is mainly suitable for determining the ET losses in the period from one irrigation to the next and during the period between two good spells of rain not heavy enough to cause percolation beyond the measurement depths.

5.4   Lysimetry

The principle behind lysimetry is to isolate a column of soil of sufficient area and depth from its surroundings and measure the weight of the isolated column. The block of soil is enclosed in a suitable container. In this either a core of soil of area and depth equal to the container, is dug out and placed in the container or the container may be back-filled i.e., the soil is dug out from the field in layers and carefully filled in a proper order to simulate the soil and bulk density profile. The instruments used to measure Evapotranspiration, ET of crops are called Lysimeters. At least 2 and possibly 3 lysimeters should be used in ET measurements. The lysimeters should be surrounded by an appreciable cropping area. The crop raised in the lysimeter must be planted, fertilized, watered and managed in the same way as the surrounding field. The surrounding of lysimeters must be so planted to avoid any non-cropped periphery around the lysimeters.

Depending on the method used to determine the change in weight of the soil-filled container, lysitmeters are classed as mechanical, mechanical cum electrical, floating, hydraulic or buoyed types. In the mechanical type the container is placed on a weighing machine. In the mechanical cum electrical type the dead load of the container is counter balanced by a lever system and the live load is recorded by means of a strain gauge load cell, the resistance of which is proportional to the live load. In the floating type, the field tank is placed in a slightly bigger container and the annular space is filled with a suitable liquid to cause flotation of the inner tank. Change in fluid level in the annular, which space gives a measure of ET, is measured. In the type, the lysimeter is supported on water filled rubber tubes hydraulic and the pressure generated is measured in a manometer filled with water and open at one end. The height of water in the manometer is measured by a float-gauge. Differences in height readings give a measure of ET. The buoyed type consists of a hydraulic lysimeter placed inside a container with the annular space filled up with a suitable liquid. Changes in level of the liquid in the annular space gives a measure of ET.

For measuring water consumption of anaerobic crops with standing water or with water table close to the soil surface, a lysimeter for provision of water table at any desired depth above or below soil is used. For aerobic crops rotated with the anaerobic ones the water table is lowered such that while the root zone is at field capacity the surface soil layers in the lysimeter dry out as in the surrounding field (Venkataraman, 1982).

Drawbacks

Daily visit to lysimeters to manually record the weight and other lysimetric parameters may lead to disturbance of field crop surrounding the lysimeter. At experimental stations, in case of irrigated crops in summer, the lysimeter field may be an isolated one with bare surroundings. In such case advection of heat from the field surrounding the lysimeter will lead to excessive values, about 60 to 70% higher, than that relating to an extensive stretch of the same crop.

5.5   Factors in Analyses of Evapotranspiration Data

5.5.1   Evaporative Losses

5.5.1.1 From bare soil

The Total Evaporable Water, TEW is the amount of water that can be lost via the soil surface from a soil column at its field capacity moisture status. TEW varies with soil types (Allen et al., 1998) being 20 mm in fine textured soils, 15 mm in medium textured soils and 10 mm in Coarse textured soils. In each soil type, TEW is equally contained in a surface and sub-surface soil layer. From a wetted bare soil column, evaporation occurs at the rate of EPA till the surface layer dries out. EPA is the potential capacity of air to desiccate moist substances and which is the resultant of a number of weather factors like solar radiation, relative humidity and wind speed. Soil moisture loss then occurs from the sub-surface soil layer at a rate controlled by soil Hydraulic Conductivity with daily evaporation equal to K $(t - t_1)^{1/2}$, where t is time of day in the declining phase and K has values 5.0, 4.0 and 3.3 respectively in fine, medium and coarse soils (Ritchie, 1972).

5.5.1.2 Under Crop Cover

With growth of the crop the soil surface gets increasingly shaded and EPA over the soil surface, EPAs progressively decreases to a steady value of 0.2 times EPA at time of ground-coverage by the crop (Ritchie and Burnett, 1971; Venkataraman et al., 1976). The latter is reached when the Leaf Area Index (LAI), which is the ratio total leaf area of the crop to the area of land surface occupied by the crop, reaches, 3.0 (Ritchie and Burnett, 1971). Even many short crops have maximum LAIs of 5.0 or more (Venkataraman and Krishnan, 1992). Thus, further increase in LAI beyond the ground-shading crop stage will not lead to any reduction in EPAs.

5.5.2 Transpiration

Under an invariant EPA i.e., constant weather conditions, the transpiratory loss per unit area of land increases with crop growth and reaches a limiting value when the crop (a) fully governs the ground in case of short crops and (b) develops a full canopy in case of tall crops. Tall crops can consume more water compared to short ones (Monteith, 1965).

5.5.3 Potential Evapotranspiration

The postulation of Potential Evapotranspiration (PET) (Thornthwaite, 1948; Penman, 1948) as the maximum amount of water that can be lost by a short, green, vegetative, ground-covering crop at any given ambient condition and the feasibility of determining PET from empirical formulations requiring use of routinely available meteorological data gave an impetus for studies on meteorological determination of water needs of crops as PET and EPA can be considered as synonyms.

5.5.4   Reference Evapotranspiration

A plethora of terms used by scientists as synonyms of PET have related to surfaces ranging from turfs to tall crops. The empirical formulations for derivation of various components of PET have also varied. In view of the above, the FAO Expert Consultation on revision of methodologies for crop water requirement (Smith et. al. 1992) had recommended, for universal adoption, an equation for computation of Reference Evapotranspiration (ET). However, it should be noted that ET refers to a short-turf of 12 cm height and that the peak water needs of agricultural crops are higher than that of turf-grass (Allen et. al, 1989) and taller crops consume more water than short ones (Monteith, 1965). Details for computation of ET from data on bright sunshine hours, temperatures, relative humidity and wind speed are set out in Appendix I.

5.5.5   Pan Evaporation

Evaporation, EP, from water filled pans, called evaporimeters, integrates the effects of all parameters that determine PET. So EP can also be considered as a measure of EPA. Pan Evaporimetry is the real-time choice for determination of (i) optimal irrigation scheduling of clear season crops (ii) net irrigation requirements of irrigated crops passing through the rainy season and (iii) provision of advisories, both for regular and supplementary irrigation, on an operational basis.

However, EP varies with material of construction, dimensions and manner of installation of evaporimeters and is liable to vitiation by visitation of birds. In India EP data are being recorded with a mesh-covered standard Class a Pan. However, energy will get advected into the evaporimeter from its sides and bottom even when the observatory yard is surrounded by crops. The quantum of this advection will be high when the observatory surroundings are bare. Pc values for converting EP data from various evaporimeters to ET under different exposure and weather conditions have been given by Allen et al. (1998). So the ratio of EP/EPA, called the Pan Coefficient, Pc can vary with seasons and regions (Venkataraman et al, 1984) and crop-wise coefficients for converting ET for peak water needs of crops will be required.

5.5.6   Relative Evapotranspiration

The ratio of ET/EPA is called Relative Evapotranspiration RET and is synonymous with the term Crop-Coefficient, Kc (Doorenbos and Pruitt, 1977; Allen et al. 1998). In the above EPA may be ET or may be calibrated, as mentioned above, in terms of EP (Venkataraman et al. 1984). The concept of RET has helped in evaluating the role of weather, crop and soil factors on water needs and water availability to crops and in agroctimalological planning of irrigation scheduling of crops and agrometeorological advising of timing and quantum of irrigation of crops as detailed below.

5.5.6.1 Temporal Field March of RET

Concurrent availability of daily lysimetric ET data and EPA can be used to plot the daily march of RET vide Fig. 1. which is also known as Crop-Coefficient Curve (Allen et al.1998). Fig.1. shows that (i) RET value, which is maximal on days following an irrigation begins to decline with days with the value before irrigation representing the transpiration component (ii) there is a gradual and a steady increase in pre-irrigation values of RET to a peak value at time of full development of the crop canopy and (iii) the peak RET value is maintained till the onset of the maturity phase.

It is the concept of RET that has brought out the fact that water needs during maturity is governed by the maturity physiology which varies amongst crops (Hattendorf et al., 1988; Venkataraman, 1985). For example RET will (a) continue to be high in case of crops like hybrids of millets, sorghum and maize that show no senescence during maturity (b) decline to a steady value for crops, like in sugarcane, that can control water uptake during maturity and (c) progressively drop to very low values at harvest stage in case of crops, like, wheat, whose foliage begin to dry up during maturity (Venkataraman, 1985). RET values also show that local varieties of Sorghum have the ability to curtail water consumption during periods of soil moisture stress (Venkataraman, 1981a).

Dates of phenological crop stages and dates and amounts of rain can be marked in the curve of temporal march of RET. Such information for the RET curve for a given crop and soil type can help determine (i) the time of commencement of active uptake of water by the crop (ii) time attainment of the ground cover stage typified by ET equalling EPA (iii) times and duration of peak water consumption (iv) peak values of RET and (v) RET value at harvest.

RET values after an irrigation will drop sharply due to soil moisture stress. The quantum of ET loss in the period between the date of irrigation and the date of sharp drop in RET values give the quantum of easily available for crop use. Such a quantity is also known as the Limiting Level of Evapotranspiration, LLEO (Ritchie et al.; 1972). Data on LLEO shows that it can vary (i) with crops in the same soil (ii) with soil types for the same crop and (iii) with level of EPA for the same crop and soil type (Venkataraman, 1981b). Thus, for a given crop the fraction of the total water available with ease for crop use will vary with EPA and soil types (Venkataranan and Krishnan, 1992). The higher the EPA and the higher the clay content of soil the lesser will be the ease of availability of soil moisture.

The time of attainment from sowing of the stages of active water uptake, ground-cover, commencement and end of peak water consumption and harvest may be specified in terms of growing degree-days above a base temperature appropriate to the crop. Such specification is extremely useful in light of the fact that though the degree-day requirements for completion of life-cycle may vary amongst cultivars, the percentage distribution amongst various phases will be the same (Venkataraman, 2003; Venkataraman et al., 2005).

5.6   Agroclimatic Planning of Irrigation Scheduling of Clear Season Crops

The advantage of agroclimatic determination of optimal irrigation scheduling is that the same can be carried out for all prospective command areas for all prospective relay crops therein. Climate at a place has a bearing on choice of crops for irrigated relay cropping. However, water needs of crops is principally weather-controlled due its direct effect on transpiratory needs of crops and indirect effects on life-duration and growth-rhythm of crops. Thus, at a given region the total water needs will nearly be the same for different sequences of relay crops of same times and duration of field-occupancy. Herein lies a great advantage of agroclimatic determination of crop water needs.

Ideally assessment of irrigation scheduling for crops has to be done on crop-climate zone basis and for the period best suited for maximal yield in each zone. Thus, delineation of homogeneous climate zones for the crop under consideration is a must. In each zone there will be contingency planning with varieties of appropriate duration for early and late start of the season. Irrigation scheduling for such early and late sown cultivars also need to be made. For details, a study on agroclimatic determination of irrigation scheduling for early, normal and late sown wheat crop in India (Venkataraman, 2004a; 2005a) for the delineated wheat-climate zones in India (Venkataraman and Rahi, 1983) may be referred to.

5.6.1   Evaporimetric Irrigation Scheduling

Computation of ET on a weekly, climatological basis for many agromet Pan Evaporation stations and on a monthly basis for a large number of synoptic meteorological stations are possible as outlined in Appendix. I. From these, Pc values can be worked out and extrapolated in time and space. The value of K_c *Pc will give values of RET in terms of EP.

As detailed above, the processes of evaporation, E and transpiration T are differently affected under a given regime of EPA. Allen et al., (1998) have suggested the use of dual coefficients of T_c for determination of transpiratory need and E_c for assessment of evaporative depletion. Use of separate coefficients of T_c and E_c is inevitable in water balance studies of rainfed crops (Venkataraman, 2005b) or for assessment of supplementary irrigation needs in the rainy season (Venkataraman, 2004b). However, for clear season irrigated crops E can be treated as a fixed fraction of 0.20 ET for a ground-covering crop and hence use of separate coefficients of T_c and E_c is not necessary for assessing ET needs of crops from ground cover stage to physiological maturity. Use of T_c and E_c would appear necessary for assessing the ET need of crops in the establishment phase from sowing to ground cover in the clear season but can be obviated through the use of the concept of Potential Crop Transpiration, PCT (Smith et al., 1992) as enumerated below.

The total water needs of a crop will depend upon its life duration and growth-rhythm. Therefore, as spelt out by Venkataraman (1995, 2004a), for evaporimetric irrigation scheduling data on (i) values of EP (ii) times and duration of crop phases from (a) sowing to ground-coverage (b) ground cover to commencement of peak water consumption (c) peak water consumption and (d) decline from peak water consumption to physiological maturity (iii) Peak RET value (iv) RET at end of maturity and (v) maximum quantum of soil moisture available with ease for crop use as appropriate to the crop and soil type under consideration are required.

Water needs in each of the above four phases can be determined as follows. When the coverage of soil by the crop is complete its transpiratory need will be 80% of PET. So in the crop establishment phase from sowing to ground-cover, the water need for transpiration will be 40% of product of cumulative pan evaporation and Pan Coefficient P_c for that period and place. To the quantum of transpiratory need, the water that will be lost due to evaporation from the soil surface has to be added. The evaporative loss will equal the number of irrigations required times the TEW for the soil type.

For the 3 stages from ground-cover to physiological maturity, the crop water need will be equal to the product of cumulative pan evaporation in the stage times the average of Crop Coefficients K_c at the end and beginning of the stage times the Pan Coefficient Pc appropriate to the place and period. Thus, for the phase from ground cover to beginning of peak water consumption K will be average and K_c Max. During peak consumption K will be K_c Max. From end of peak water consumption to physiological maturity, K will be the average of K_c Max and K_c at physiological maturity.

5.6.2   Irrigation Interval

For the 3 stages from ground cover to physiological maturity, the irrigation interval will be equal to cumulative pan evaporation divided the quantum of easily available soil moisture for crop use, LLEO, appropriate to the crop, soil and evaporative power of air under consideration as detailed earlier. In the crop establishment phase both leaf area development, transpiratory need and quantum of soil moisture that can be foraged by the crop stand increase with crop growth and irrigation must be done to ensure that the crop does not have a moisture stress. Heavy pre-sowing irrigation is not enough to guarantee freedom from soil moisture stress. Detailed examination of this problem (Venkataraman, 1995) shows that the irrigation interval in the establishment phase will remain invariant and half as much as that required during the peak consumption stage. Data on optimum irrigation interval to meet peak crop water demands can be had as mentioned above and are also available from field irrigation trials.

For farmers getting water on assured dates, ease of reservoir operations and command area management and bulk mono-cropping, an uniform interval of irrigation is preferable. Irrigation is to be given ahead of crop-water consumption. So the last irrigation and irrigations during crop establishment become critical. Once phasic irrigation needs are determined as above for a crop and a location, from considerations of root zone moisture capacities, appropriate to the crop and soil type, the timings and interval of irrigation and quantum of water applied at each irrigation can easily be worked out.

5.6.3 Assessment of Supplementary Irrigation Needs

Long duration irrigated crops frequently pass through a rainy season. A crop like sugarcane can pass through two rainy seasons. In such cases, it is necessary to assess the extent to which rainfall will meet the crop water needs for proper planning for provision of supplementary irrigation in the rainy season for various crops in different areas. Such an assessment involves determination of Effective Rainfall, ERR, which is the quantum of rain getting past the evaporative barrier and contributing to the crop's water needs.

As mentioned above evaporative loss in the crop phase from sowing to ground coverage of an irrigated crop in the clear season can be taken as equal to the number of irrigations times the Total Evaporable Water, TEW appropriate to the soil type and equal to 10, 15 and 20 mm in coarse, medium and fine textured soils. In TEW, Readily Available water, RAW equal to 50% of TEW will be lost at rate of EPA while the other 50% of TEW will be lost, in the rainy season, as vapour transport at rates of 1.50, 1.25 and 1.00 in fine, medium and coarse textured soils respectively (Venkataraman, 2005 b). Vapour transport will set in only after total loss of RAW. Wetting of soil by rains amounts to random irrigation both regarding amounts and intervals. Thus, for assessment of ERR and of the timing and quantum of supplementary irrigation, it is necessary to calculate separately the transpiratory consumption of the crop and evaporative depletion of rain by the use respectively of the Transpiratory Coefficient, T_c and Evaporative Coefiicient E_c (Allen et al., 1998). Both T_c and E_c will relate to EPA. The age when the crop enters the rainy season has a bearing on the selection of time unit for agrometeorological analyses. If the rainy period sets in after ground-cover stage of the crop, weekly periods can be used for rainfall budgeting. However, if the rainy season occurs in the crop establishment phase daily rainfall budgeting has to be done till ground-cover crop stage. Thus, for long duration sugarcane crop daily rainfall budgeting in the first rainy season and weekly rainfall budgeting in the second rainy season will be required.

Any crop would have been irrigated to field capacity moisture status prior to onset of rains. The period to start rainfall budgeting will be the day of irrigation prior to onset of rains. On that day values of E_c and T_c must be chosen. For established crop, weekly values E_c will be 0.20 while T_c will relate to peak water consumption. In the crop establishment phase E_c values have to be progressively decreased to reach a value of 0.20 at ground-cover stage while T_c values have to be progressively increased to peak value at full canopy. To start with TEW will be maximal appropriate to the soil type and root zone moisture will be at field capacity moisture status. In both weekly and daily computations, transpiratory need will be $T_c \times$ EPA. In daily computations RAW equal to 50% of TEW will be lost at $E_c \times$ EPA. As mentioned above, the remaining quantum of TEW can be allowed for at a steady daily rate of 1.50, 1.25 and 1.00 mm per day in fine, medium and coarse textured soils respectively.

Cumulative ERR must exceed TEW for soil moisture accretion in the crop root zone. Budgeting of cumulative ERR in terms of transpiratory need can be done independently except to take account of ERR infusions. Thus, supplementary irrigation may be needed even when the soil surface is moist. In irrigation, 75% depletion of root zone moisture is taken as the limit before irrigation is due. However, for assessment of timing and quantum of supplementary irrigation the limit for depletion of root zone moisture may be taken as 2/3 of root zone moisture quantum. A sample computation of Supplementary Water Need in the Kharif season for a short crop of initial Root Zone Moisture, RZM capacity of 80 mm growing in fine textured soil with initial TEW of 20 mm, entering the rainy season after germination and growing under an EPA 5.0 mm per day is given vide Table 5.1A.

Table 5.1A Sample computation of supplementary irrigation needs in kharif season

Crop Age, Days	E E_{Max} mm	T mm	TEW mm	RZM mm	RRR mm	E Act. mm	ERR mm	SWN mm
9			20.0	80.0	0	0	0	
10	4.0	1.0	16.0	79.0	0	4.0	0	0
11	3.9	1.1	12.1	77.9	0	3.9	0	0
12	3.7	1.3	9.4	76.6	0	2.7	0	0
13	3.5	1.5	12.9	75.1	7	3.5	0	0
14	3.4	1.6	11.5	73.5	2	3.4	0	0
15	3.3	1.7	9.4	71.8	0	2.1	0	0
16	3.1	1.9	7.9	69.9	0	1.5	0	0
17	2.9	2.1	15.0	67.8	10	2.9	0	0
18	2.8	2.2	14.2	65.6	2	2.8	0	0
19	2.7	2.3	11.7	63.3	0	2.5	0	0
20	2.5	2.5	9.6	60.8	0	2.1	0	0
21	2.3	2.7	18.3	58.1	11	2.3	0	0
22	2.2	2.8	16.1	55.3	0	2.2	0	0
23	2.1	2.9	15.0	52.4	1	2.1	0	0
24	1.9	3.1	20.0	49.3	7	1.9	0.1	0
25	1.7	3.3	18.3	46.0	0	1.7	0	0
26	1.6	3.4	16.	42.4	0	1.6	1.2	0
27	1.5	3.5	20.0	38.9	6	1.5	2.7	0
28	1.3	3.7	20.0	43.1	10	1.3	8.7	0
29	1.1	3.9	20.0	45.1	7	1.1	5.9	0
30	1.0	4.0	19.0	41.1	0	1.0	0	40.0
			20.0	80.0				
5th Week	7.0	28.0	20.0	65.0	20	7.0	13.0	0.0
6th "	7.0	28.0	20.0	47.0	17	7.0	10.0	0.0
7th "	7.0	28.0	20.0	24.0	12	7.0	5.0	56.0
			20.0	80.0				
8th "	7.0	28.0	20.0	80.0	35	7.0	28.0	0.0
9th "	7.0	28.0	20.0	55.0	10	7.0	3.0	0.0
10th "	7.0	28.0	20.0	31.0	11	7.0	4.0	49.0

In Table 5.1A. daily values of Evaporation E and Transpiration T equal to $E_c \times EPA$ and $T_c \times EPA$ respectively are given from date of last irrigation to ground-cover stage with daily values of E_c and T_c

chosen as explained above. RRR and ERR are rainfall and Effective Rainfall respectively. In Table 5.1A, T relates to transpiration from crop with no soil moisture stress, E_{Max} is Potential Evaporation from a shaded crop and E Act is the actual daily evaporation. In calculating E Act, in the time left in the day after cumulative evaporative loss of 10 mm, E is allowed for at fixed rate of 1.5 mm per day. RZM and TEW are progressively reduced by the T and E Act of next date. RR of date must be added to TEW of date and deduced from E Act of next day to get TEW of date. TEW in excess of 20 mm will be ERR and TEW on that day will be 20.0 mm. ERR of date gets added to RZ of previous date. Supplementary irrigation to raise the soil column to field capacity status is to be given when RZ gets close to 30 mm and/or when rains are about to start. It is seen from Table 5.1A that out of a rainfall of 63 mm only 14.5 mm contribute to the crop's water need. Again, supplementary irrigation is required even with a wet soil surface. After crop establishment supplementary fortnightly irrigations of about 40 mm each are required.

Now use of daily values for computation is taxing. Use of T_c for weekly periods even during crop establishment will not affect the transpiratory needs. Since a crop will be irrigated to field capacity status before irrigation, average weekly values of E_c will only marginally affect ERR as shown in Table 5.1B for same data as at IA but for weekly periods.

Table 5.1B Sample computation of SWN during crop establishment using weekly data

Days	E_c	T_c	E	T	TEW	RZM	RRR	ERR	SWN
10-16	0.7	0.3	25	10	4.0	70	9.0	0.0	0.0
17-23	0.5	0.5	17	17	11.0	53	24.0	0.0	0.0
24-30	0.3	0.7	10	25	20.0	39	30.0	11.0	40.0

In dryland cropping 50% is an acceptable risk level. In rainy season EPA is 1.1 times EP (Venkataraman, 2001). For agroclimatic determinations of SWN, start computations in the week prior to the one when Minimum Assured Rainfall, MAR at 50% probability exceeds 1.1 times EP. Use weekly values of MAR at 50% probability values for RRR. Use appropriate values of E_c and T_c corresponding to stage of crop at time of pre-rainy season irrigation.

5.7 Water Use-Efficiency

The term Water Use-Efficiency WUE is used by irrigation engineers to indicate the fraction of water let out from the reservoirs that is used in crop production at field level. To the agronomist WUE is the economic crop yield produced per unit amount of water used in crop production. There is no physiological basis for relating crop yields to their water requirements as both high and low yielding cultivars of the same crop of same crop-life duration and growth-rhythm and exposed to the same weather will have the same water need. Since yield of a given crop in a given season is directly related to its field-life duration, ratio of water used to the field-life duration of a given common cultivar will be a good measure for comparing areal variations in WUEs of the crop cultivar. The indications are that while WUE of a crop cultivar at any given location will not be much affected by earliness or lateness of sowing compared to normally sown crops, WUE efficiency for a given crop can vary across locations widely (Venkataraman, 2005a).

5.7.1 Net Income per Irrigated Hectare per Unit Quantum of Water used

Crop-wise delineated homogenous climate zones of irrigated crops will throw up a cafeteria of feasible cropping practices from which it is possible to select for irrigated relay cropping, crops and their cultivars and their sequencing for maximal productivity in a given region. Because of agroecological reasons, even for a given crop its rotational crops cannot be the same in all regions. The water requirements for relay crops will mainly be determined by time of field-occupancy of crops. The criteria of water-use efficiency applied to single crops will not be valid for determining water use efficiencies of irrigated relay cropping systems. The use of the concept of Net Income per Hectare (Swaminathan, 2006) in judging performance of varieties and technologies has been made. However, the prevalent Minimum Support Prices, MSP are not weighted for variations in crop productivity due to the climatic influence and charges for irrigation water are based on area irrigated and not on quantum of water supplied. The above concept and that of WUE can, however, be combined to evolve the criterion of productivity per hectare per unit of irrigation water used. The above criterion will be a valuable comparative yardstick in deciding location-specific cropping pattern for maximisation of gross crop productivity, WUE and net income for farmers.

5.8   Savings in Irrigation Water: Agronomic Measures

5.8.1 Systematic Rice Intensification

Nowhere is the need for saving irrigation water more apparent and urgent than in lowland irrigated rice, where the crop is raised on puddled soil with standing water. Puddling of soil needs enormous amounts of water. Percolation loss from standing water is inevitable and seepage loss is also significant. Thus, depending on soil and seasons, the water needs of puddled rice can be 2 to 3 times of that of an aerobic irrigated crop.

The methodology of Systematic Rice Intensification SRI (Satyanaryana et al, 2007) has come to be adopted in more than 20 countries, as a low water-use, low-cost, environmentally benign and sustainable alternative system to that of flooded rice, especially in water scarce regions. SRI has had an accidental origin in 1983 in Madagascar (Laulane, 1993). The success of SRI arises from the physiological fact that young rice seedlings retain their potential for formation of tillers if they are transplanted before the start of the 4th phyllochron (Stoop et al, 1992), i.e., before 15 days of age in the tropics.

In the SRI methodology, 8 to 12 day old seedlings, with just two leaves and with seed, soil and roots intact are transplanted gently in a muddy field within 12 hours of removal from the nursery at a depth of 1 to 2 cm, singly in a square pattern of 25 cm × 25 cm with the roots lying horizontally in the moist beds. Till the roots get established, a thin layer of water is let into the field at night and water is drained away in the morning. Afterwards the soil is kept moist but not saturated and irrigation is resorted to when the surface cracks up. Presence of thin depth of water in the reproductive crop phase is not mandatory and alternate wetting and drying can be resorted to. Weeding, by a rotary hoe, is done at 10 days interval from the 10th day of transplanting till the crop canopy closes. Weeds are incorporated in the soil. Only organic manures or compost is used in SRI.

The water management practices of SRI avoid the pitfalls of flooded rice culture. Compared to conventional flooded rice, SRI saves 40 to 50% of irrigation water. Yield increases of 20 to 50% under

SRI over flooded rice has been reported in many countries. The SRI practices even when not exclusively used, such as SRI irrigation with conventional planting or SRI planting with conventional irrigation, enhances the yield level of rice (Horie et al, 2005). One could, thus, surmise that with the same quantum of water as is being used in puddled rice, the crop output can easily be doubled or trebled.

Yields under SRI are critically dependant on strict adherence to the recommended geometry of planting, population density and crop-water management practices. Thus, direct seeding is not advised (Satyanarayana, 2005) and transplantation is mandatory. Because of the above SRI entails high labour cost as right from starting of nursery till harvest skilled labour for more days though for fewer hours per day is required. To practise the procedures recommended for irrigation, an on-the-tap system of irrigation is required. Provision of such a system on a large scale will be a daunting task. On a command area basis SRI has to be grown on a block system with standardized timing of operations, and uniform cultural cum irrigation practices. Organic system of fertilisation to the SRI crop is quite costly.

5.8.2 Irrigated Aerobic Rice

Irrigated, Aerobic Rice does not require puddling of soil and of standing water and various methods are adopted for provision of water for the crop. Savings in water in aerobic rice vis-a-vis the flooded rice arise from savings in water needed for land preparation and absence of seepage and percolation. Studies on comparative water use and yields of aerobic, irrigated rice with flooded rice in the dry season show that the savings in irrigation water in aerobic rice culture more than offset the reduction in yield vis-à-vis flooded rice. The above can be translated into a larger area under rice for the same quantum of water, especially in situations where the farmers do not have access to enough water to grow flooded lowland rice. Thus, from the point of view of efficient use of water in increasing rice production, adoption of aerobic irrigated rice culture is called for.

5.8.3 Sugarcane and Summer Cotton

Replacement of spring-planted one year Sugarcane crop by June planted 18 months' crop just ahead of the rainy season, as is done in Peninsular India, will not only increase productive crop-life duration but also reduce net irrigation needs. Sugarcane regulates its water uptake during maturity. So if maturity period is adjusted to occur in warm weather irrigation water can be saved. A similar approach can save water in case of summer planted cotton.

5.8.4   Micro-Irrigation

The use of micro-irrigation, also called drip-irrigation, has been mooted to save irrigation water in crop culture by directly delivering water to crop roots and thus avoiding the wasteful evaporative losses. Micro-irrigation is a dire necessity when crops have to be established in the warm weather period using groundwater. Micro-irrigation Use-Efficiency, MIUE may be defined in terms of the water used in micro-irrigation to what would have been required to be used as surface application. The above discussions, on evaporation from bare soil and soil surfaces of incomplete and complete crop covers, show that use-efficiency of micro-irrigation (a) will be maximal and minimal respectively during crop establishment in warm weather and for a ground covering winter crop and (b) can be agrometeorologically assessed for given soil types, crop conditions and level of EPA in terms of normal surface irrigations required as follows:

Sowing to crop establishment: MIUE = $n*E_s/\{(k*EPA*N) +(n . E_s)\}$

In the above n= number of surface irrigations normally required in the phase; E_s = Total evaporable water and equal to 10, 15 and 20 mm in coarse, medium and fine soils; k = 0.40 for short crops and 0.50 for tall crops; N is the number of days of the phase and EPA is the average EPA in the phase in mm/day.

Ground-coverage to commencement of maturity: MIUE (Micro-Irrigation Use Efficiency). Numerator will be the lesser of the two quantities, $n*E_s$ or $N*0.20$ EPA. The denominator will be (K*EPA* N) + (value of numerator.) In the above E_s is Total Evaporable Water qual to 10, 15 and 20 mm in coarse, medium and fine soils. k = 0.80 for short crops and 1.0 for tall crops, N and EPA are respectively number of days and average EPA in mm/day in the phase.

Commencement of maturity to harvest: MIUE. For short and tall crops not showing physiological senesence during maturity, MIUE will be the same as for the phase ground- coverage to commencement of maturity. For other crops, MIUE can be computed using vales of 0.40 and 0.50 for short and tall crops respectively as for the phase ground-coverage to commencement of maturity.

5.8.5   Sprinkler Irrigation

Sprinkler irrigation tries to simulate provisioning of water from rainfall and is the best means for irrigating crops in undulating terrain. When crop canopy is complete the water collected by leaf foliage may be conducted to the root zone by stem flow to circumvent evaporative depletion. Ideally the sprinkler systems should be operated nocturnally to avoid evaporation of water sprays from the sprinklers. A perusal of literature shows that water use efficiency of sprinkler irrigation is 20% less than that of drip irrigation. The spacing required to ensure full areal coverage, time-duration of operations to avoid soil moisture stress to crops and use of mechanisms to ensure proper water droplet size in the sprays makes installation of sprinkler systems very costly compared to drip irrigation. Thus, the system can be used when the value of the crop to be saved is high. For example yield of Tea is markedly depressed in the absence of pre-rainy season rains (Sen et al., 1966) and tea requires supplementary irrigation in the post-rainy season (Biswas and Sanyal, 1971). Sprinkler irrigation for Tea may be worthwhile. For other crops, in undulating terrain, the system can be employed on a hiring basis only.

5.9   Groundwater

Rainfall not lost as surface run-off, not evaporated, not lost through the wick-like action of transpiration gets stored in the soil. No loss of water will occur from soil depths beyond the root zone of crops once they are raised to field capacity. Thus, the amount of carry over soil moisture in excess of that required to raise root zone of crops to field capacity can be taken as groundwater storage. Thus, groundwater accretion from rainfall is the last of the parameters in rainfall budgeting for assessment of groundwater storage. However, since any rainfall budgeting will address all the entities mentioned above, groundwater storage will accrue as a byproduct of rainfall budgeting for crop use. Whether all the assessed quantities will ultimately show up as groundwater storage depends on soil geology.

Thus, agroclimatic analyses and considerations of soil geology will assist in the fraction of rainfall that is likely to get stored as groundwater. Restricting use of groundwater (a) to average annual recharge from rains and (b) using the same within the above limits for critical irrigations for (i) saving rainfed crops during droughts (ii) marginally extending the maturity periods of an irrigated relay crop nearing warm weather and (iii) drip and spray irrigations will improve water use-efficiency of groundwater. Use of groundwater as a sole source of irrigation to increase gross irrigated acreage and in warm weather period must be discouraged. Free or subsidised electricity to farmers must be restricted to operation of open dug wells and/or low power pump sets.

Maximisation of Rainfall Usage

Contrary to popular beliefs and as pertinently observed by the Indian National Commission for Farmers in its first report in 2004 (NCF, 2004), water use efficiency of crops is 10 to 30% lower under irrigation compared to non-irrigated conditions. Unlike irrigation rainfall is free. By (i) maximizing the direct us e of rainfall in crop production (ii) harvesting surface runoff from rains through construction of agroclimatically designed (Srivastava, 2001), cost-effective (Pandey et al., 2005) On-Farm Reservoirs, OFRs (Bhatnagar et al., 1996) and re-using the water so collected in drought situations and (iii) augmenting recharge of groundwater and strategically using the same to combat drought or extending the maturity periods of crops, the cost of water used in dryland crop production can be kept minimal. Simultaneously if the farmers with irrigated holdings are charged for irrigation on the basis of amounts of water supplied, differences in net income per unit area under dryfarming and irrigation can be much narrowed down. The above aspects are more fully and comprehensively covered in a later chapter on "Dryland Agrometeorology".

References

Allen, R.G.; Jensen, M.E.; Wright, J.L. and Burman, R.D. 1989. Operational estimates of reference evapotarspiration. Agron. Jl. 81: 650-662.

Allen, R.G., Periera, L.S., Raes, D. and Smith, M. 1998. Crop evapotranspiration: Guidelines for computing crop water requirements. Irrigation and Drainage paper No. 56, Food and Agriculture Organisation (FAO), 300 pp.

Bhatnagar, P.R.; Srivastava, R.C. and Bhatnagar, V.K. 1996. Management of runoff stored in small tanks for transplanted rice production in mid-hills of North-West Himalaya. Agric. Water Management. 30: 107-118.

Biswas, A.K. and Sanyal, D.K. 1971. Rainfall and Irrigation. Part I. Two and A Bud. 18: 12-19.

Bowen, I.S. 1926. The ratio of heat losses by conduction and by evaporation from any water surface. Phys. Rev. 27: 779-787.

Deacon, E.L. and Swinbank, W.C. 1956. Comparison between momentum and water vapour transfer. Arid Zone Research. Climatology and Microclimatology. Proc. UNESCO Symposium of Arid Zone Climatology, Canberra, pp 38-47.

Doorenbos, J. and Pruitt, W.O. 1977. Crop water requirements. FAO Irrigation and Drainage paper No.24. 144 pp.

Gadgil, S. and Seshagiri Rao, P.R. 2000. Farming strategies for a variable climate: A challenge. Current Sci., 78: 1203-1215.

Garg, K.N. and Hasan, Q, 2007. Alarming scarcity of water in India. Curr. Sci. 93: 932-041.

Grant, R.F. et al. 1999. Crop water relations under different CO_2 and irrigation: Testing of ecosystems with the Free Air CO_2 Enrichment (FACE) experiment. Agric. and Forest Meteorol. 95: 27-51.

Hattendorf, M.J.; Redelfs, M.S.; Amos, B.; Stone, L.R. and Gwin, R.E. 1988. Comparative water use characteristics of six row crops. Agron. Jl. 80: 80-85.

Horie et al. 2005. Can yields of lowland rice resume the increases that they showed in the 1980s? Plant Prod. Sci. 8: 257-272.

Laulane, H. 1993. Le systeme de riziculture intensive malgache. Tropcultura, 11: 110-114.

Monteith, J.L. 1965. Evaporation and Environment. Proc. XIXth Symposium of Society for Experimental Biology. Cambridge University Press. 205-234 pp.

Muchinda, M.R. and Venkataraman, S. 1988. Influence of temperature and moisture on maturation periods of rainfed maize. Productive Farming, September 1988, 20-25.

National Commission on Farmers, NCF, 2004. First Report - Serving farmers and farming. Private communication. 28 pp.

Pandey, P.K.; Panda, S.N. and Pholane, L.P. 2005. Modeling for maximizing precipitation utilization in rainfed agriculture in Eastern India. Bull. Nat. Inst. Ecol. 16: 113-120.

Penman, H.L. 1948. Natural evaporation from open water, bare soil and grass Proc. Royal Soc. London, Series A 193, 120-146.

Raman, C.R.V.; Venkataraman, S. and Krishnamurthy, V. 1973. Dew over India and its contribution to winter-crop water balance. Agricl. Meteorol. 11: 17-35.

Ritchie, J.T. 1972. Model for predicting evaporation from a row crop with incomplete crop cover. Water Resources Research. 8: 1204-1213.

Ritchie, J.T. and Burnett, E. 1971. Dryland evaporative flux in a sub-humid climate. II. Plant Influences. Agronomy JI, 63: 56-62.

Ritchie, J.T.; Burnett, E. and Henderson, R.C. 1972. Dryland evaporative flux in a sub-humid climate. III. Soil water influences. Agronomy. Jl. 64: 168-178.

Ritchie, J.T. and Ne Smith, D.S. 1991. Temperature and Crop Development. In: R.J. Hank and J.T. Ritchie (Editors). Modeling Plant and Soil Systems. Agronomy Series No. 31. Am. Soc. Agron. Madison. Wisconsin, U.S.A. Chapter 3, pp 31-54.

Satyanarayana, A. 2005. System of rice intensification – The need of the hour. Spl. Lecture, 34th Research and Extension Advisory Council Meeting, Agricultural University, Hyderabad, India. Personal Communication.

Satyanarayana, A.; Thiyagarajan, T.M. and Uphoff. N. 2007. Opportunities for water saving with higher yield from the system of rice intensification. Irrigation Sci. 25: 99-115.

Sen, A.R.; Biswas, A.K. and Sanyal, D.K. 1966. The influence of climatic factors on yield of Tea in the Assam valley. Applied Meteorol. 5: 789-800.

Smith. M. et al. 1992. Report Expert consultation on revision of FAO methodologies for crop water requirements, FAO, 60 pp.

Srivastava, R.C. 2001. Methodology for design of water harvesting system for high rainfall areas. Agric. Water Management, 47: 37-53.

Stoop, W.; Uphoff, N. and Kassam, 2002. A review of agricultural research issues raised by the System of Rice Intensification (SRI) from Madagascar: Opportunities for improving farming systems for resource-poor farmers. Agric. Syst. 71: 249-274.

Swaminathan, M.S. 1968. Genetic manipulation of productivity per day. Spl. Lecture. ICAR Symp. on "Cropping Patterns in India".

Swaminathan, M.S. 2006. 2006-07: Year of agricultural renewal. Public lecture, 93rd Indian Science Congress, Hyderabad, January 2006. 32 pp. Private communication.

Thornthwaite, C.W. 1948. An approach toward a rational classification of climate. Geographical Review 38: 55-94.

Venkataraman, S. 1981a. Lysimetric observations on moisture accretion for and use by M-35-1 jowar at Solapur. Jl. Maharashtra Agric. Universities. 6: 36-40.

Venkataraman, S. 1981b. Some Preliminary Observations on the Evapotranspiration of Rabi Groundnut at Hyderabad, The Andhra Agric. Jl. 28: 12-17.

Venkataraman S. 1982. A Volumetric Lysimeter System for Use with Puddled Rice and its rotational crops. Mausam, 33: 91-94.

Venkataraman, S. 1985. Agrometeorology as a link in technology transfer for the stabilization of crop outputs. International Jl. Ecol. and Environ. Sci. 11: 91-103.

Venkataraman, S. 1995. Agrometeorological determination of the optimal distribution of total water requirement of crops. Internat. Jl. Ecol. and Environmental Sci. 21: 251-261.

Venkataraman, S. 2001. Agrometeorological anticipation of yields of rainfed crops. Ind. Jl. Environ. & Ecoplan. 5: 135-144.

Venkataraman, S. 2003. Thermal relations of phenology of crops: A case study for Maize. Ind. Jl. Environ. & Ecoplan. 7: 291-293.

Venkataraman, S. 2004a. Agroclimatic determination of water needs of crops - A case study for wheat in India. Ind. Jl. Environ. & Ecoplan. 8: 565-568.

Venkataraman, S. 2004b. A crop-weather model for assessing net irrigation requirements of sugarcane in India under varied planting times and field-life durations. The Ekologia.2: 31-36.

Venkataraman, S. 2005a. Climatic influences on water use-efficiencies in irrigated wheat in India. Jl. Cur. Sci. 7: 277-280.

Venkataraman, S, 2005b. A crop-weather model for water balance studies of Kharif crops through rainfall budgeting. Ind. Jl, Environ, & Ecoplan. 10: 285-290.

Venkataraman, S. and Krishnan, A. 1992. Crops and Weather. Pub. ICAR, 586 pp.

Venkataraman, S. and Rahi, A.K. 1983. Influence of temperature climatology on productivity of wheat crop in India. Mausam, 34: 81-84.

Venkataraman, S.; Subba Rao, K. and Raghava Rao, P. 1976. A preliminary lysimetric study on the evapotranspiration of sugarcane at Anakapalle IMD Pre Pub. Sci. Rept.No.76/13, 12 pp.

Venkataraman, S.; Subba Rao, K. and Jilani, Y. 1984. A comparative study of the climatological estimation of potential evapotranspiration. Mausam. 35: 171-174.

Venkataraman, S.; Kashyapi, A. and Das, H.P. 2005. Phasic distribution of total heat units in wheat cultivars in India. Mausam. 56: 499-500.

Appendix I

Reference Evapotranspiration, ETo.

The FAO (Smith et al 1992) recommended equation, for universal adoption, for computation of Reference Evapotranspiration, ETo can be written as follows, namely:

$$\text{ETo} = \frac{0.01707\,\Delta(R_n - G) + \left[\gamma * 100 * U2 * (e_a - e_d)/[3*(T+273)]\right]}{\Delta + \gamma(1 + 0.0944\,U2)} \quad (1)$$

Dividing both numerator and denominator by γ, Eq.1 can be written as:

$$\text{ETo} = \frac{(\Delta/\gamma) * 0.01707\,(R_n - G) + \left[100 * U2 * (e_a - e_d)/[3*(T+273)]\right]}{(\Delta/\gamma) + (1 + 0.0944\,U2)} \quad (2)$$

In the above: ETo is Reference Evapotranspiration in mm/day;

R_n is Net Radiation at the crop surface in cal/cm²/day.

G is Soil Heat Flux in cal/cm²/day;

T is mean air temperature in degrees centigrade;

U2 is mean wind speed at 2 meters height in kilometers per hour. ea and ed are saturation vapour pressure in mm at mean air and dew point temperatures

$e_a - e_d$ is saturation vapour pressure deficit in mm

Δ is the slope of vapour pressure versus temperature curve

γ, is the Psychometrics constant, which can vary with station height

In the above (i) correction factors for reduction wind from measurement height to 2 meters can be had from Table 2.9 of Annex 2 of FAO publication of Allen et al. (1998) and (ii) values of ea and ed can be had Vide Table 4.4 of chapter 4. Values of Δ/γ, the dimensionless ratio given in Tabsle 1, for temperatures from 6 to 45 degrees centigrade, refer to a station height of 300 meters and can be used with an error of plus or minus 3% for stations up to 600 m. For station heights other than 300 meters the values have to be reduced or increased respectively for stations of higher or lower heights by 1% for every 100 meters of difference in heights.

Table 5.1 Value of the dimensionless ratio Δ/γ

T ᵒC	Δ/γ	T ᵒC	Δ/γ	T ᵒC	Δ/γ	T ᵒC	Δ/γ	T ᵒC	Δ/γ
06	1.000	14	1.600	22	2.477	30	3.739	38	5.508
07	1.061	15	1.692	23	2.615	31	3.939	39	5.769
08	1.123	16	1.785	24	2.754	32	4.139	40	6.046
09	1.200	17	1.892	25	2.908	33	4.339	41	6.339
10	1.261	18	2.000	26	3.061	34	4.554	42	6.631
11	1.339	19	2.108	27	3.215	35	4.785	43	6.939
12	1.415	20	2.231	28	3.385	36	5.015	44	7.246
13	1.508	21	2.354	29	3.554	37	5.261	45	7.585

In equation 2, one can neglect G over a 24-hour period. In the above Net Radiation R_n = Non-reflected Solar Radiation plus non-reflected Long wave Radiation from the Atmosphere minus the long wave back-radiation from the crop cover.

Measured values of Solar Radiation Rs are available for many locations. Methodology for derivation of Rs from data on bright hours of sunshine have been detailed in Chapter 4. The reflection Coefficient, Albedo, for a green ground-covering crop cover is 0.23. (Allen et al. 1998). The albedo and emissivity of surfaces for long wave radiation are independent of the nature of the surface and have values of 0.03 and 0.97 respectively. The radiative temperature of a freely transpiring, vegetative, green and ground-covering crop will be equal to that of mean air temperature, Ta (Linacre1968). Therefore, for a crop as above, long-wave back radiation at mean air temperature Vide Table 5 of chapter 4 may be used. Methodology of computation of Long Wave Radiation from the Atmosphere. Rlw↓ has been detailed in Chapter 4. Climatological monthly ratios of net to back radiation by BTs[4],where Ts in mean air temperature in degrees absolute and B is the classical Stefan-Boltzmann constant will give Rlw↓(Venkataraman,1977). As mentioned in Chapter 4, on the basis of values of net long-wave radiation as computed above, calibrated constants in place of the ones advocated by Allent et al.(1998) for computing Net Long Wave Radiation can be and needs to be worked out.

Regarding Vapour Pressure, if only dry and wet bulb temperatures are available, Dew Point temperature and hence e_d can be obtained from temperature and wet bulb depression from psychometric tables appropriate to height of the station. If only maximum and minimum values of temperature and relative humidity are available, vapour pressure will be average of (i) saturation vapour pressure at maximum temperature times minimum relative humidity and (ii) saturation vapour pressure at minimum temperature times maximum relative humidity.

References

Allen, R.G., Periera, L.S., Raes, D. and Smith, M. 1998. Crop evapotranspiration: Guidelines for computing crop water requirements. Irrigation and Drainage paper No. 56, Food Agriculture Organisation (FAO), 300 pp.

Linacre, E.T. 1968. Estimating the net-radiation flux. Agric. Meteorol. 5: 49-63.

Smith. M. et al. 1992. Report, Expert consultation on revision of FAO methodologies for crop water requirements, FAO, 60 pp.

Venkataraman, S. 1977. Evaluation of mean daily net longwave flux using pyrgeometric data. Mausam, 28: 402-4.

CHAPTER 6

Dry Farming Agrometeorology

The important rainfall features affecting dryland crops are (i) the onset and withdrawal of rains and (ii) the temporal distribution of rainfall over short periods of time with reference to recharge of soil moisture for meeting the water needs of the crop. For the meteorologist, onset and withdrawal of rains relate to wind circulation patterns in the upper atmosphere and associated changes relating to perceptible water content and differences in wet and dry bulb temperatures of air in the lower air layers. However, for dryland crops one has to consider commencement and end of Agriculturally Significant Rains (ASRs). Rains that will enable start of land preparation for sowing with a reserve of moisture in the seedling root zone for germination and seedling establishment will be start of ASR. Rains that will not contribute to crop water needs will be end of ASR. End of ASR is not end of the crop period as a crop can thrive on moisture available in the root zone when rains cease. For a given temporal distribution of rainfall, the ideal crop is the one that will have a soil moisture stress-free vegetative phase and will complete its full life without any moisture stress of its maturity period with the help of root zone moisture available at end of rains.

6.1 Dryland Cropping Period (DCP)

The times and duration of a crop such as the above can be termed the Dryland Cropping Period (DCP). The week or the dekad (10 days) are the accepted time-units for agrometeorological studies, over such short periods, rainfall amounts show very high inter-year variations at a place for a given week/dekad. Thus, DCP will, a priori, vary amongst the years at a place. Even on a climatological basis the duration of the rainy period shows inter-regional variations. Thus, DCP will also show areal variations. Therefore, assessment of DCP has to be done on a station-wise, year-wise basis using long-series rainfall data, Frequencies of occurrence of DCPs of various durations thus assessed can assist in planning crop cultivars and cropping systems suited to local rainfall climatology so as to minimise the effects of adverse rainfall years and maximise the benefits from good rainfall years.

6.2   Rainfall Probabilities

Year by year analyses of short-period, long series, rainfall data is a daunting task. The year to year variations of short period (weekly/Dekadal) rainfall at a place are very high. The mathematical breakthrough achieved by way of the Incomplete Gamma Distribution Model (Thom, 1966) had helped in dealing with such variations on a probability basis (Mooley, 1973) i.e., find the Minimum Assured Rainfall (MAR) in a given week/dekad at a given probability level or find the level of probability of realising a specified quantum of MAR in a given week/dekad. Many meteorological services have published Rainfall Probability data for weekly or dekadal periods for a very large net-work of representative stations. For example, the India Meteorological Department (IMD) has published data on weekly MAR amounts at probability levels of 30, 40, 50, 60 and 70% for 1950 taluka level stations on a state-wise basis (IMD, 1995). The rainfall probability data are of great use in assessment of DCP for a given rainfall distribution as detailed below.

6.3   Review of Methodology

The two most widely used parameters in assessment of DCP are MAR at various probability percentage levels and PET. Use of monthly periods (Hargreaves, 1971) for the delineation of DCP is not suitable, as features of short-period rainfall would get masked out over a month. Rainfall probability levels of 75% (Hargreaves, 1971) and 70% (Muchinda and Venkataraman, 1986) are too high a level of expectancy in dryland agriculture, where 50% probability can be deemed to be an acceptable risk level.

Dryland crops are sown with weekly rainfall of at least 20 mm (Victor et al., 1988), preferably accumulated over 2 consecutive days with a reserve of 15 mm in the root zone to facilitate germination (Victor et al., 1996). Now rainfall has to fill the evaporative zone to field capacity before moisture accretion for crop use can occur. Moisture in the evaporative zone can be used to prepare the soil to the required tilth for sowing. EPA is about 4.5 mm per day in the dryland farming tracts in the rainy season (Cocheme and Franquin, 1967). The transpiratory need of the crop in early stages is 0.20 times EPA. So a reserve of 20 mm can help the crop to survive a rainless post-sowing spell of 3 weeks. The duration of rainless, post-sowing survival will depend vitally on the quantum of moisture available for crop use at time of sowing. The capacity of the evaporative zone is 20, 15 and 10 mm respectively in fine , medium and coarse textured soils (Allen et al., 1998).   Thus, the postulations that start of the rainy season as the one when rainfall minus PET reaches 20 mm (Victor et al., 1996) and that over weekly periods rainfall must exceed PET (Fitzpatrixk and Nix, 1967; Venkataraman, 1979) for soil moisture accretion are valid. The latter stipulation would also apply to the dekadal period.

Use of low ratios of rainfall: PET of 0.34 (Hargreaves, 1971) and 0.30 (Frere and Popov, 1979) will lead to a non-realistic (i) earlier start of the cropping period and (ii) higher carry over soil moisture from rains and   result in gross overestimation of DCP. Use of ratios of rainfall: PET of 50% (Cocheme and Franquin, 1967; Brown and Cocheme, 1969) and criterion of 20 mm of MAR at 50% probability level (Venkataraman, 1981) would all over estimate DCP.

The suggestion to accumulate 50 mm rainfall backwardly from date of maturity and then proceed forward towards maturity date until average daily rainfall becomes less than 2.5 mm /day and take that date as the date of final rain has been mooted (Victor et al., 1996). Dryland crops mature in a rainless period. In fact rains during the late maturity period will be yield-depressing. For crops that show little physiological senescence during maturity the soil moisture needs will nearly be equal to EPA in the maturity period.   Even for crops that dry up at harvest, the ET need in the drying phase will be 70% of EPA. Thus, what is of importance is the time of ceassation of ASRs and quantum of soil moisture available at end of ASRs.

6.4   Criteria for Delineation of DCP

The following criteria have been suggested (Venkataraman, 2001) to delineate times and duration of DCP from weekly rainfall probability data of MAR and the same can be adapted for dekadal periods :

(a) Week with MAR of 30 mm or more followed by week of at least 25 mm MAR will be the start of the ASR period. For dekadal period start of ASR will be MAR of 45 mm followed by dekad with MAR of 35 mm. The stipulation of 2 successive weeks/dekads is to eliminate the problem of assigning false starts to pre-rainy thunderstorm rains.

(b) The week/dekad from which MAR sharply falls off below 15 mm and 25 mm respectively will be the end of the ASR period.

(c) MAR in excess of EPA will be the Soil Moisture Carry-Over, SMC.

(d) SMC must be used to make good the deficiency in specified MAR in earlier weeks/dekads.

(e) The maximum quantum of SMC that can be carried over after budgeting as at D above must be restricted to a limiting value appropriate to the crop and soil type

(f) The quantum of SMC at the end of ASR divided by EPA× K_c, where K_c is the crop co-efficient for the maturity phase of the crop, will give the number of days of crop growth period after end of ASRs.

6.5   Delineation and Utility of Rainfall Zonation

Once times and duration of DCP is determined for a large number of staions on a 50% probability basis, delineation of rainfall zones, with intra-regional homogeny and inter-regional differences with reference to the starting and ending of ASRs, quantum of soil moisture accretion at end of the ASR period and times and durtion of DCP, can be carried out. Such a zonation can help in (i) ascertaining the areal extent of applicability of findings of dryland research stations (ii) proper selection of minimum number of representative stations for agrometeorological monitoring of the performance of rainfed crops (iii) detailed micro-level agroclimatological analyses for (a) planning of on-farm-reservoirs for harvest and re-use of surface run-off from rains (b) initiation of measures for augmentation of groundwater (c) selection of crops/varieties and field practices to suit local rainfall regimes and DCPs and (iv) adoption of uniform cropping practices in a zone.

6.6   Rainfall Harvesting and Re-use

Due to rainfall vagaries, dryland crops frequently face the risk of severe moisture stress leading to drying up of the crop. Digging out a portion of the main field to collect surface run-off from rains in an On Farm Reservoir (OFR) and using the OFR storages to save the crop from drought has been mooted and is in practice. OFRs are kept free of crops and hence constitute non-productive areas in a field. For determining the optimum fraction of field to be set apart for OFRs (Bhatnagar et. al., 1996) it is necessary to carry out rainfall budgeting to arive at the temporal march of the quantum of surface runoff from rains on the one hand and times and extent of likely moisture stress on a climatological basis. Use of rainfall budgeting models for designing OFRs is feasible (Srivastava, 2001; Pandey et al., 2005).

In conventional rainfall budgeting, when rainfall exceeds EPA the excess amount is deemed usable subsequently by the crop and any surplus exceeding the maximal field capacity status of the crop is deemed as not available for crop use. However, in real-time, rainfalls can runoff from soil surfaces even before the evaporative process begins. The amount of run-off depends on the moisture status of the soil, infiltration capacity of the land and rainfall intensity. In rainfall budgeting, assessment of runoff from rains has to be done on a daily basis and is the most difficult. Techniques for determining surface run-off from rains, such as the widely accepted one by U.S. Deptt. of Agriculture (USDA, 1972), involve daily rainfall analyses. Assessment of daily runoff from for a large number of station-years will be a daunting task and hence have to be restricted to a station in each of the delineated rainfall zones.

Experimental data on surface run-off from rains under different soil covers are available at a number of soil conservation research centres. Agroclimatically determined values have to be first calibrated against appropriate experimental data. Such validated estimates will then apply to the entire rainfall zone in which the experimental centre is located. The next step will be to derive the surface run-off estimates from more readily available rainfall data so that the times and quantum of surface run-off from rains can be computed on a climatological basis for many more stations.

Runoff collections in OFR will be subject to evaporation, percolation and seepage losses. OFRs will be small in size and being sunk are largely free of play of wind on the water surface. Thus, evaporation can be considerably reduced by use of long-chain alcohols to form mono-molecular film over water surfaces. Seepage and percolation can be prevented by lining the bottom and sides of the OFR by low-density polythene (LPDE) sheets. The above cost money and it is necessary to assess the cost-effectiveness of the measures in terms of value of crop saved (Pandey et al., 2005).

6.7   Rainfed Rice

Rice crop is associated with plentiful and enormous waste of water. A discussion of rice as a rainfed crop may, therefore, seem out of place in dryland cropping. However, many regions during the rainy season regularly experience stormy rains, are subject to riverine floods and are characterized by heavy rains of 100 mm per week or so over an extended period, Rice is the only suitable crop that can be grown in such places. Areas lying above the flood plain constitute the upland, rainfed rice ecosystem, in which the rice crop is raised in non-bunded fields. A rainfall regime of 100 mm per month for 4 consecutive months is considered suitable for upland rainfed rice culture; there are two types of flood prone areas, namely Deep-Water Rice Areas and Tidal Wetlands. The former are found in the lowland, deltaic areas of rivers where water accumulates for 30 days or more to depths of half to three meters in the rainy season. Tidal wetlands are in coastal areas subject to risk of seawater intrusion on account of storm surges. Rainfed lowland rice, rainfed upland rice and rice from flood prone areas account for 17%, 4% and 4% respectively of global rice production.

6.7.1   Rainfed Lowland Rice

Rainfall is the only water source for the Rainfed lowland rice ecosystem. Rainfed lowland rice is raised in places where surface irrigation is not available and there is risk of inundation of fields from rains for significant periods of time in the crop season. In this rice is grown in bunded fields with overflow arrangements to ensure that the depth of standing water remains less than 50 cm over a period of

10 consecutive days. Rainfed lowland system is characterized by (i) uncertainty in time of start of the crop season and (ii) intermittent ponding, saturation, wetting and drying of the soil in a random manner.

6.7.2 Rainfall Budgeting for Design of OFRs for Lowland Rice

The rainfall budgeting procedure required for design of an OFR for lowland rice (Pandey et al., 2005) has some features distinctly different from that for an aerobic crop. Effective rainfall for land preparation is high and about 200 plus or minus 50 mm depending on soil type. Lack of standing water reduces yield. So rainfall need is that required to maintain a specified water level. The limiting values of carry over moisture will be determined by the saturated moisture capacity of root zone and not field capacity moisture. The phase before water gets ponded is the unsaturated phase while the saturated phase begins when ponding of water becomes feasible. As and when it becomes possible to pond the water, allowance has to be made for seepage and percolation losses, which are highly variable (Srivastava, 1996). Seepage and percolation losses can be treated as a combined loss (Wickham and Singh, 1978) and can be allowed for at a fixed daily rate depending on soil type. ET losses/requirements in the saturated phase will equal EPA. Surface runoff will occur whenever rainfall intensity exceeds the infiltration rate of the soil and is to be assessed on a daily basis in a manner similar to that of aerobic crops by the SCS curve number procedure of USDA (1972).

6.7.2.1 Methodology

(i) Cumulate on a daily basis differences between rainfall, RR and EPA. Take negative values as zero. Set limiting value of such cumulations as equal to Saturated Soil Moisture Content (SSMC) of root zone appropriate to the soil type.

(ii) Assign values in excess of SSMC as Depths of Water available for Ponding (DWP).

(iii) The time Σ DWP reaches the value for land preparation is the start of the saturated phase.

(iv) In the saturated phase, starting with a given depth of water, add daily rainfall to depth of standing water and minus EPA, seepage and percolation losses and run-off and assign excess if any to depth of water available for ponding.

(v) The time when such cumulations lead to nil depth of water is the time of onset of water deficiency and the water need will be equal to the desired depth of water level, D.

Now cumulative D will give the extent of deficiency of rainfall in meeting the rice water needs. Cumulative runoff will give the quantum of harvestable rain. The ratio between cumulative D and harvestable run-off will give the fraction of area of the field to be set apart as OFR. Application of the above methodology for a large number of years at a location will give, on a probability basis, the quantum of supplementary irrigation need and the quantum of surface runoff that can be harvested. The two parameters can help in deciding the fraction of main-field that should be set apart for constructing the OFR.

6.7.3   Rice cum Fish Culture

The Irrigated Lowland Rice Ecosystem, in which rice raised in puddled soil with standing water, is an obvious habitat for fish. Thus, the question of augmenting the income of rice farmers through rice cum fish culture had been mooted by FAO in the late 1940s. The idea could not catch on till late 1970s (Ghosh and Saha, 1978). The percentage of rice area under rice cum fish culture is only 0.05 % in India, about 4% in China and nearly 40% in Egypt. The highest area, nearly 3 million ha and the highest fraction, 32%, of total area under rice cum fish is found in Thailand.

The requirements of water depth, temperature, pH, oxygen and water-turbidity for fish and rice for optimal performance are quite different. Also water level cannot be allowed to fall below a specified minimum while fish stocks are present in the rice field. Chemical control of pests, diseases and weeds of rice will harm fish. When grown together fish damage the rice crop. The above calls for raising of fish concurrently with rice but stocked separately in a fish-pen forming a separate area of the field (Bhatnagar et al., 1996).

Rice cum fish culture under the irrigated lowland rice ecosystem would require additional quantum of irrigation water as fish pens would require initially substantially higher amounts of water. SRI or aerobic irrigated rice systems do not permit fish culture. In deep-water rice areas and tidal rice wetlands, stocked fish may escape from the rice fields due to overflow of bunded fields by floodwaters. Thus, rainfed lowlands emerge as the only suitable system for rice cum fish culture.

(Pandey et al., 2005) have carried out a detailed study on the economics of fish production from an optimally sized OFR. The study indicates that the additional fish production in OFR will more than compensate for the loss of rice crop from the OFR area and give an additional income for the farmer practicing rice cum fish culture in fields with intermittent ponding and drying.

6.8   Crop Drought Climatology

A climatological assessment of the extent of proneness to crop droughts in various regions of the dryfarming tract is required by planners for according priorities in organising programmes for alleviation of the sufferings of dryland farmers. This calls for adoption of criteria for classification of different intensities of drought and the assessment of frequnecies of occurence of different classes of crop drought at a given location as per criteria laid down. A linear relationship between yield of crops and their seasonal evapotranspiration, ET above a threshold value has been observed (Musick, 1994; Zhang and Oweiss, 1999). As transpiratory consumption of water by crops constitute a major fraction of seasonal ET, it follows that for optimal growth and yield. Transpiration of crops must be maintained at the potential rate i.e., the crops must remain free of soil moisture stress. Therefore, for rainfed crops a "Crop Drought" may be deemed to occur whenever the rainfall derived root zone moisture becomes insufficient to meet the needs of the crop for "Potential Transpiration".

Amongst Meteorological, Hydrological and Agricultural droughts, the last is the most difficult to assess. Hydrological droughts call for evaluation of seasonal total rainfall received in catchment areas of reservoirs vis-a-vis the normal amounts. In meteorological droughts the criteria are based on percentage departures of weekly rainfall amounts from the normal at a given station/region. Crop drought assessments call for analyses of rianfall on a daily basis in the crop establishment phase and on a weekly basis thereafter for determining the temporal march of accrued root zone moisture and its sufficiency to meet the phasic demands of crop for potential transpiration.

The times and duration of a crop raised at a place is based on the local experience of rainfall climatology of the location. The manifestation of drought effects on a crop is influenced by soil factors. Thus, crop drought assessments can only be comparative in time at a place but not across areas.   They have therefore, to be done on a station-wise basis and at a station on a year-wise basis. Delineation of dryland farming zones as detailed above will assist in carrying out such areal assessments on a reasonably representaive basis.

6.8.1   Review of Methodology

Criteria for meteorological droughts have no agricultural significance. For example even a 50% departure of rainfall from normal in a high rainfall zone need not mean crop drought and in fact may be beneficial to crops due to non-occurrence of floods. Similarly even with above normal rains in a semi-arid area, crop droughts can occur. Derivation of a mean drought index from climatological averages and computation of extent of dviation of yearly drougt indices from the mean (Subrahmanyam et al., 1965; Palmer, 1965) have the same drawbacks as the approach of departures from normal rainfall. Analyses relating to a crop of DCP that appears most appropriate to the location in a delineated dryland farming zone would amount to begging the question. Local assessment of the percentage frequency of occurence of crop life durations free of soil moisture stress should be the over-riding consideration.

6.8.1.1 Water Requirements Satisfaction Index (WRSI)

Frere and Popov (1979) had postulated the concept of Water Requirement Satiation Index (WRSI) to estimate yields in terms of percentge of the potential yield (Popov, 1984) of the same crop i.e., a crop not subject to any soil moisture under the same local environment. Computation of WRSI involes the use of weekly/dekadal values of rainfall and PET in the following formula:

$$SM_i = SM_{i-1} RR - WN$$

wherein SM_{i-1} is the soil moisture at start of the week/dekad, SM_i is the soil moisture at the end of the week/dekad, RR is the weekly/dekadal rainfall and WN is the water need of the crop for the week/dekad and equal to $K_c \times PET$ where K_c is the crop coefficient appropriate to the crop and its stage and age. In this the deficit refers to the shortfall of rainfall in meeting the water need of the crop after taking into account the moisture availability at start of the week/dekad. When RR is greater than WN, the excess after making good the soil moisture deficiencies of earlier weeks/dekads is restricted to a limiting value appropriate to the crop and soil type.

WRSI is assumed to be 100 at the beginning of the crop season as crops are sown with adequate moisture in the soil. The value remains at 100 for succeding weeks/dekads until a water stress occurs. In that case the percentage ratio of RR-WN/WN is calculated from the WRSI value at the end of the preceding week/dekad. The calculation is continued till the end of the crop period. When RR is greater than WN but occurs in less than 3 days, WRSI is reduced by 2.1 points to account for influence of water logging on the crop yield.

The welcome features of the above approach are (i) its simplicity (ii) cognizance of (a) need for yearly analysis and (b) crop factors like life-duration and develeopmental rhythm and soil attributes relating to ease and maximal quatum of availability of soil moisture (iii) computation of WRSI on a weekly or dekadal cum cumulative phases for allowing determination of the times, duration, intensity and growth stage of occurence of the moisture stress. The last feature is of importance as crops can recover from moderate, non-prolonged moisture stresses occuring in growth stages of the crops not very sensitive to moisture stress. As crop yields will be more affected if soil moisture stress occurs during sensitive growth stages, WRSI during such stages can be given appropriare weightage (Doorenbos and Kasam, 1979). In many models developed to relate seasonal moisture stress indices it has been noted that weighted indices correlate better with crop yields (Shaw, 1974, 1977; Sudar et al., 1981).

Good correlations between WRSI and crop yields have been reported for (a) groundnut at (i) Bambey, Senegal (Frere and Popov, 1979; Popov, 1984) (ii) Hydrerabad and Anantpur, Andhra Pradesh, India (Raji Reddy et al., 2002; Yogeshwar Rao et al., 1990) (iii) Rajkot, Rajasthan, India (Srivastava et al., 1989) (b) Pearl Millet, Sunflower, Castor, and Pigeon Pea at Hyderabad, Andhra Pradesh, India (Victor et al., 1991). Computation of WRSI does not account for the influence of biotic and abiotic stresses on crop yields. So relationship, if any found, will only hold forth in years free of incidence of pests and diseases and hazardous weather phenomena. WRSI is a measure of percentage of potential yield realisable at a location. In yield analyses, one has to account for technology trend. So, for examination of relationship between WRSI and crop yield, one must use most recent data at the location relating to potential yield realised.

6.8.1.2 Transpiratory Satiation Index

Crops are generally sown after ascertaining availability of moisture in the seeding zone of the crop. Because of the small transpiratory need of the crop during germination and establishment, the crop can survive without rain for a fort night or so. However, for such a recharge of rainfall to occur over bare soil, the rainfall has to get past the formidable barriers of evaporative losses from the surface layer and satiation of the surface layer to its moisture capacity. The Crop Coefficient, K_c used in WRSI computations is the sum of the Transpiratory Coefficient T_c and the Evaporative Coefficient E_c. K_c values are small to start with. However, taking rainfall need of the crop as equal to PET×K_c will grossly underestimate the rainfall deficit.

6.8.1.3 Rainfall Budgeting For Assessment of TSI

Procedures and formats for computation of the temporal march of TSI under a given rainfall distribution for a given crop, growing in a given soil have been detailed by Venkataraman, (2005). The methodology takes into account the process of evaporation from soil surfaces detailed in Chapter 5 and the following features.

As mentioned earlier, with crop growth, E_c decreases, slowly to a steady value of 0, 20 at the ground-cover stage of the crop, Due to the above the evaporative component can be taken as a fixed fraction of ET from the 5th week of crop growth with E_c values of (i) 0.35 and 0.25 for the 5th and 6th week of crop growth respectively and (ii) 0.20 from the 7th week onwards (Venkataraman, 2003). Over bare soil and in the first month of crop growth irregular wetting of the soil column with varying amounts of rainfall renders assessment of evaporative depletion of rainfall on a daily basis mandatory.

Assessment of the evaporative depletion of rainfall from the surface evaporative layer of the soil is straightforward. Assessment of the evaporative loss from the sub-surface evaporative layer can be simplified by taking into account the fact that the values of Hydraulic conductivity are 5.0, 3.3 and 4.0 respectively for fine, coarse and medium textured soils (Ritchie, 1972) and the daily evaporative loss from the sub-surface layer will range from (a) 2.0 to 1.0 mm in fine textured soils (b) 1.60 to 0.80 in medium textured soils and (c) 1.35 to 0.67 in coarse tetured soils (Ritchie, 1972). Thus, the average daily evaporative loss can be taken as 1.50, 1.25 and 1.0 mm respectively in fine, medium and coarse soils without any serious error in the estimation of total quantum of evaporative loss from the sub-surface evaporative layer (Venkataraman, 2005).

Rainfall budgeting for a ground-covering crop, i. e., from the 5th week or 4th dekad of crop growth is quite simple. Rainfall budgeting over bare soil, though to be done on a daily basis is not complicated. The stage from sowing to ground-cover involves daily budgeting and more complex analysis. The ground-cover stage of the crop is a physical and not physiological stage and depends on initial rate of seeding and population density of seedlings. However, errors arising from the assumption of 30 day duration for this stage will not materially affect computation of TSI. Thus, rainfall budgeting for assessment of temporal march TSI has to be done for 3 stages, namely (i) bare soil up to sowing (ii) sowing to ground-coverage and (iii) ground-coverage to physiological maturity.

6.8.2 Basic Data Needs

The data required for assessment of TSI are: (i) Daily Rainfall and EPA (ii) Soil Type (iii) Crop and times and duration of its phases, namely: (a) Bare soil to sowing/planting (b) Sowing/Planting to ground-cover; (c) Ground-cover to physiological maturity (iv) daily values of E_c in the first month of crop growth (v) weekly values of T_c appropriate to the crop and its age and life duration (Vide Table 6.4) and (vi) limiting values of carry-over moisture appropriate to crop and soil type (vide Table 6.5).

6.8.3 Formats, Legends, Procedures and Sample Computations

As rainfall budgeting has to be done on a daily basis up to one month of crop growth, a 3 layer, 3 stage formats, similar to that of Frere and Popov (1986) has been proposed by Venkataraman (2005). The formats along with explanatory legends and procedures for assessment of soil moisture accretion and stress Days and sample computations are given and detailed. Table 6.1 relates to assessing, over bare soil, accretion of soil moisture in the potential root zone of Hybrid Jowar crop of 120 days duration growing in a fine-texture soil. Computation of Stress Days (SDs), in first Month of Crop Growth is detailed in Table 6.2. Computation of Stress Days from Second Month Crop age to Harvest is detailed in Table 6.3.

Table 6.1 Daily budgeting over bare soil of fine texture; soil moisture capacity of surface and sub-surface evaporative layer is 10 mm each; maximum root zone moisture is 80 mm.

1	Day	A	B	C	D	E	F	G	H	I	J	K	L	M	N
2.	RRR	0.0	0.0	16.5	8.0	3.5	0.0	0.0	10.5	11.5	13.0	0.0	0.0	21.5	7.5
3.	EPA	5.0	5.0	5.0	5.0	5.0	5.0	5.0	4.5	4.5	4.5	4.5	4.5	4.5	4.5
4.	E_c	1.0	1.0	1.0	1.0	1.0	1.0	1.0	1.0	1.0	1.0	1.0	1.0	1.0	1.0
5.	ΣEL_1	0.0	0.0	10.0	10.0	8.5	3.5	0.0	6.0	10.0	10.0	1.0	1.0	10.0	10.0
6.	ΣEL_2	0.0	0.0	1.5	4.5	4.5	4.5	4.5	4.5	7.5	10.0	10.0	10.0	10.0	10.0
7.	RZ	0.0	0.0	0.0	0.0	0.0	0.0	0.0	0.0	0.0	6.0	0.0	0.0	8.0	3.0
8.	ΣRZ	0.0	0.0	0.0	0.0	0.0	0.0	0.0	0.0	0.0	6.0	6.0	6.0	14.0	17.0

Legend of New Terms:

Days over bare soil are designated in alphabets.

RRR is daily rainfall in mm; ΣEL_1 & ΣEL_2 are Soil moisture accretions in surface & sub-surface evaporative layers respectively; RZ and Σ RZ are daily and cumulative root zone moisture accretion.

Procedure:

Start computations on day when RRR exceeds EPA.

Cumulate RRR minus EPA values. Take negative values also as zero.

When sum exceeds surface layer capacity but is less than TEW assign excess to sub-surface layer.

If the sum is more than TEW assign excess to RZ. When surface layer is dry, allow for daily evaporation from sub-surface layer at rates as appropriate to soil type.

Week when Σ RZ becomes 15 or more is the sowing week.

135

Computation of Stress Days (SDs), in first Month of Crop Growth

Table 6.2 Hybrid jowar of 120 days duration; Soil type: Fine textured; Soil moisture capacity of surface and sub-surface evaporation layer is 10 mm each; Maximum root zone capacity is 80 mm.

	Day	1	2	3	4	5	6	7	8	9	10	11	12	13	14
1	Day	1	2	3	4	5	6	7	8	9	10	11	12	13	14
2	RRR	0.0	0.0	9.5	8.8	17.5	0.0	0.0	0.0	0.0	0.0	0.0	0.0	8.1	7.6
3	EPA	4.0	4.0	4.0	4.0	4.0	4.0	4.0	4.5	4.5	4.5	4.0	4.0	4.0	5.0
4	E_c	0.99	0.97	0.96	0.94	0.93	0.91	0.90	0.87	0.84	0.81	0.79	0.76	0.73	0.70
5	ΣEL_1	6.0	2.1	7.8	10.0	10.0	6.4	2.8	0.0	0.0	0.0	0.0	0.0	5.2	9.3
6	ΣEL_2	10.0	10.0	10.0	10.0	10.0	10.0	10.0	8.5	7.0	5.5	4.0	2.5	2.5	2.5
7	RZ	0.0	0.0	0.0	2.8	13.8	0.0	0.0	0.0	0.0	0.0	0.0	0.0	0.0	0.0
8	T_c	0.01	0.03	0.04	0.06	0.07	0.09	0.10	0.13	0.16	0.19	0.21	0.24	0.27	0.30
9	PT	0.04	0.1	0.2	0.2	0.3	0.4	0.4	0.60	0.70	0.9	0.8	1.0	1.1	1.5
10	ΣRZ	17.0	16.6	16.2	18.4	31.8	31.3	30.7	29.0	29.0	28.0	27.0	25.9	24.8	22.2
11	AT	0.04	0.4	0.4	0.4	0.4	0.5	0.6	0.90	0.90	1.0	1.0	1.1	1.2	1.6
12	SDs	0.0	0.0	0.0	0.0	0.0	0.0	0.0	0.0	0.0	0.0	0.0	0.0	0.0	0.0

Table 6.2 Contd....

1	Day	15	16	17	18	19	20	21	22	23	24	25	26	27	28
2	RRR	11.6	0.0	0.0	7.4	15.9	8.8	0.0	0.0	0.0	15.4	20.5	16.4	0.0	0.0
3	EPA	5.0	5.0	4.0	4.0	4.0	4.5	4.5	4.0	5.0	5.0	4.0	4.0	4.5	4.5
4	E_c	0.67	0.64	0.61	0.59	0.56	0.53	0.50	0.49	0.47	0.46	0.44	0.43	0.41	0.40
5	ΣEL1	10.0	6.8	4.4	9.4	10.0	10.0	7.7	5.7	3.4	10.0	10.0	10.0	8.8	7.0
6	ΣEL2	10.0	10.0	10.0	10.0	10.0	10.0	10.0	10.0	10.0	10.0	10.0	10.0	10.0	10.0
7	RZ	0.0	0.0	0.0	0.0	13.1	6.4	0.0	0.0	6.5	6.5	18.7	14.7	0.0	0.0
8	T_c	0.33	0.36	0.39	0.41	0.44	0.47	0.50	0.51	0.54	0.54	0.56	0.57	0.59	0.60
9	PT	1.7	1.8	1.6	1.7	1.8	2.1	2.3	2.0	2.7	2.7	2.2	2.3	2.7	2.7
10	ΣRZ	20.4	18.6	17.0	15.3	26.6	30.9	28.6	26.6	23.7	27.7	44.2	56.6	53.9	51.8
11	AT	1.7	1.8	1.6	1.7	1.8	2.1	2.3	2.0	2.7	2.7	2.2	2.3	2.7	2.7
12	SDs	0.0	0.0	0.0	0.0	0.0	0.0	0.0	0.0	0.0	0.0	0.0	0.0	0.0	0.0

Legend of New Terms:

Days during crop period are numbered numerically and serially.

PT is the soil moisture need for Potential Transpiration in mm/day and equal to EPA × T_c

AT is actual amount of water transpired by the crop in mm/day.

SD is stress days equal to (1-AT/PT) when AT < PT.

Procedure:

For entries in Σ EL1, Σ EL2 and RZ, procedure is same as for bare soil but using values of EPs = E_c × EPA.

Add RZ of date to ΣRZ of previous date and minus from PT and enter as ΣRZ of date.

When ΣRZ is positive enter AT = PT.

Enter negative or zero values of Σ RZ as Zero.

When Σ RZ is zero, enter AT = RZ of date plus ΣRZ of previous date.

Enter SD = 1- (AT/PT) when AT< PT.

137

Computation of Stress Days from Second Month of Crop age to Harvest

Table 6.3 Weekly rainfall budgeting from steady-state evaporation to harvest.

1	Week	5	6	7	8	9	10	11	12	13	14	15	16	17
2	IRZ	51.8	42.4	41.4	64.4	47.2	28.4	21.0	24.2	15.2	6.2	0.0	0.0	0.0
3	RRR	25.6	32.0	53.0	15.0	17.0	26.0	31.2	21.0	23.0	21.2	9.0	4.0	0.0
4	EPA	35.0	34.0	30.0	32.0	35.0	34.0	28.0	30.0	32.0	35.0	30.0	29.0	33.0
5	E_c	0.35	0.25	0.20	0.20	0.20	0.20	0.20	0.20	0.20	0.20	0.20	0.20	0.20
6	ERR	13.3	24.5	47.0	8.4	10.0	19.8	15.0	15.0	16.6	14.2	3.0	0.0	0.0
7	T_c	0.65	0.75	0.80	0.80	0.80	0.80	0.80	0.80	0.80	0.80	0.80	0.80	0.80
8	PT	22.7	25.5	24.0	25.6	28.8	27.2	22.0	24.0	25.6	28.0	24.0	23.2	26.4
9	FRZ	42.4	41.4	64.4	47.2	28.4	21.0	24.2	15.2	6.2	0.0	0.0	0.0	0.0
10	AT	22.7	25.5	24.0	25.6	28.8	27.2	22.4	24.0	25.6	20.4	3.0	0.0	0.0
11	SDs	0.0	0.0	0.0	0.0	0.0	0.0	0.0	0.0	0.0	2.0	6.0	7.0	7.0

Legend:

In the above entries against ERR, EPA, PT, AT, RZ and ΣRZ and SDs would refer to total weekly values. IRZ and FRZ are the amount of root zone moisture in mm at beginning and end of week respectively.

Procedure: Enter IRZ as equal to ΣRZ on day 28 of crop period

Enter FRZ= IRZ+ ERR – PT.

If FRZ is positive enter AT=PT.

If FRZ is zero enter AT=FRZ of previous week + ERR of week.

Enter SDs=[1-(AT/PT)] × 7.

Assessment of Intensity of Crop Drought

The computed stress days in a week give a measure of extent of satiation of the transpiratory needs of crops. For any given period the cumulative stress days expressed as a fraction of the duration of the period in days gives a measure of crop drought. The time of incidence of stress days will be indicated in terms of age of the crop in weeks from sowing and can be translated into occurrence of crop-growth stage from concurrent data on crop phenology. While moisture requirements for optimal growth may remain the same for many crops, the influence of soil moisture stress varies widely amongst crops and in the same crop, in the stage in which soil moisture stress occurs. Thus, in assessing drought incidence, stress days should be suitably weighted for given crops and their growth stages (Doorenbos and Kassam, 1979).

6.8.4 Description of Crop Drought

6.8.4.1 Intensity and Areal Spread

The total number of stress days after weighting expressed as a percentage of field-life duration of the crop will be the Transpiratory Satiation Index (TSI). Values of TSI (i) greater than 0.95 (ii) 0.85 to 0.95 (iii) 0.75 to 0.84 (iv) 0.65 to 0.74 (v) 0.55 to 0.64 (vi) 0.45 to 0.54 and (vii) less than 0.45 will relate respectively to Nil, Light, Lightly-Moderate, Moderate, Moderately Severe, Severe and very Severe incidence of crop drought.

When percentage of stations in an area covered by a particular class of drought is (i) > 90% (ii) 66 to 90% (iii) 50 to 65% (iv) 33 to 50% and (v) < 33%, description of areal spread will respectively be Widespread, Fairly Widespread, Scattered, Local and Isolated.

6.9 Crop Drought Maps

After station-wise assessment of intensity of incidence of crop drought, each district must be assessed for the areal spread of the most predominate type of crop drought.

Crop-Drought maps should depict district-wise crop drought situation in a state.

State-wise drought maps need to be integrated to produce a crop droughts position map similar to the weekly of rainfall map of IMD.

There should be two crop drought maps. In this, one map should depict the crop drought situation arising in the current week based on TSI values for that week while another map depicting the drought situation as of date ending with the week and based the TSI value for the period from sowing to end of the week should also be presented.

A hypothetical Drought Scenario map for state of Maharashtra in India which encompasses a variety of rainfall regimes is presented at Fig. 6.1.

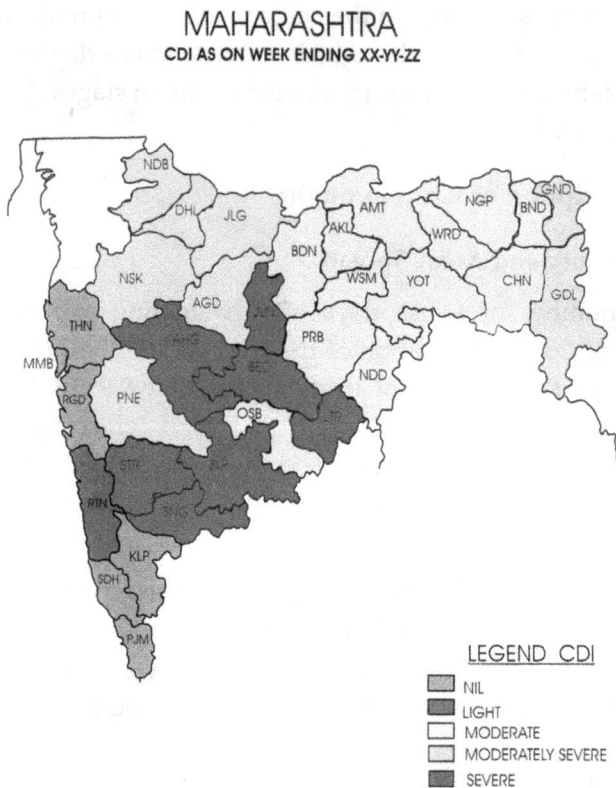

Fig. 6.1

Table 6.4 Values of T_c for some rainfed crops of varying maturity durations

Crop	Life Weeks														Crop Age in Weeks									
		1	2	3	4	5	6	7	8	9	10	11	12	13	14	15	16	17	18	19	20	21	22	23
Hybrid Maize	14	0.1	0.3	0.45	0.6	0.7	0.8	0.85	0.9	1.0	1.0	1.0	0.9	0.7	0.5									
	17	0.1	0.3	0.45	0.6	0.7	0.8	0.85	0.9	1.0	1.0	1.0	1.0	1.0	1.0	0.85	0.7	0.5						
	20	0.1	0.3	0.45	0.6	0.7	0.8	0.85	0.9	1.0	1.0	1.0	1.0	1.0	1.0	1.0	0.95	0.9	0.7	0.6	0.5			
Hybrid Millets Sorghum	14	0.1	0.3	0.45	0.6	0.7	0.8	0.8	0.8	0.8	0.8	0.8	0.8	0.8	0.8									
	17	0.1	0.3	0.45	0.6	0.7	0.8	0.8	0.8	0.8	0.8	0.8	0.8	0.8	0.8	0.8	0.8	0.8						
	20	0.1	0.3	0.45	0.6	0.7	0.8	0.8	0.8	0.8	0.8	0.8	0.8	0.8	0.8	0.8	0.8	0.8	0.8	0.8	0.8			
Local Sorghum	17	0.1	0.2	0.3	0.4	0.5	0.6	0.7	0.8	0.8	0.8	0.8	0.7	0.6	0.5	0.5	0.4	0.35						
	20	0.1	0.2	0.3	0.4	0.5	0.6	0.7	0.8	0.8	0.8	0.8	0.8	0.75	0.7	0.6	0.55	0.5	0.45	0.4	0.35			
	23	0.1	0.2	0.3	0.4	0.5	0.6	0.7	0.8	0.8	0.8	0.8	0.8	0.8	0.8	0.75	0.7	0.7	0.6	0.55	0.5	0.4	0.4	0.35
Peanut	14	0.1	0.2	0.4	0.6	0.7	0.8	0.9	0.9	0.9	0.9	0.9	0.9	0.8	0.75									
	17	0.1	0.25	0.4	0.6	0.7	0.8	0.9	0.9	0.9	0.9	0.9	0.9	0.9	0.9	0.9	0.8	0.75						
	20	0.1	0.2	0.4	0.6	0.7	0.8	0.9	0.9	0.9	0.9	0.9	0.9	0.9	0.9	0.9	0.9	0.9	0.8	0.7	0.7			
Safflower	14	0.1	0.15	0.25	0.3	0.5	0.6	0.7	0.8	0.85	0.9	0.7	0.6	0.4	0.2									
	17	0.1	0.15	0.25	0.3	0.4	0.6	0.6	0.7	0.75	0.8	0.9	0.8	0.7	0.6	0.5	0.3	0.2						
	20	0.1	0.15	0.25	0.3	0.4	0.5	0.5	0.7	0.7	0.7	0.8	0.8	0.9	0.9	0.8	0.7	0.6	0.4	0.3	0.2			
	23	0.1	0.2	0.2	0.3	0.4	0.45	0.5	0.6	0.65	0.7	0.8	0.8	0.85	0.9	0.9	0.9	0.9	0.7	0.6	0.5	0.4	0.3	0.2

Table 6.5 Generalized estimates of easily available moisture, LLEO (mm)

Crop	Fine Soil	Medium Soil	Coarse Soil
Hybrid Maize	115	115	75
Hybrid Sorghum	105	105	70
Local Sorghum	80	80	50
Hybrid Pearl Millet	115	100	65
Hybrid Finger Millet	90	80	50
Non-winter Small Grains	110	100	60
Safflower	90	90	60
Groundnut	100	105	65

References

Allen, R.R.; Pereira, L.S.; Raes, D. and Smith, M. 1998. Crop evapotranspiration-Guidelines for computing crop water requirements. Food and Agriculture Organisation (FAO), Irrigation and Drainage Paper. 56, 300 pp.

Bhatnagar, P.R.; Srivastava, R.C. and Bhatnagar, V.K. 1996. Management of runoff stored in small tanks for transplanted rice production in mid-hills of North-West Himalaya. Agric. Water Management. 30: 107-118.

Brown, L.H. and Cocheme, J. 1969. A study of the agroclimatology of the highlands of Eastern Africa. Technical Note 125, World. Met. Org. Geneva. 197 pp.

Cocheme. J. and Franquin, P. 1967. An agro-climatological survey of a semi-arid area in Africa. Tech. Note 86. World. Met. Org., (WMO), Geneva. 136 pp.

Doorenbos, J. and Kasam, A.H. 1979. Yield response to water. Irrigation and Drainage Paper No. 24. Food and Agricultural Organisation (FAO), Rome. 144 pp.

Fitzpatrick, E.A. and Nix, H.R. 1967. A model for simulating soil moisture regimes in alternating fallow-crop systems. Agric. Meteorol. 6: 303-319.

Frere. M. and Popov. G.F. 1979. Agrometeorological crop monitoring and forecasting. Irrigation and Drainage paper. 17. FAO, Rome. 64 pp.

Frere, M. and Popov, G.F. 1986. Early agrometeorological yield assessment. Plant Production and Protection Paper, 73, FAO, Rome, 154 pp.

Ghosh, A. and Saha, S.K. 1978. Scope for paddy-cum-fish culture in India. Tropical Ecology and Development. 2: 1009-1015.

Hargreaves, G.H. 1971. Precipitation dependability and potential for agricultural production in North East Brazil. EBMRAPA and Utah State Univ. Publication. No.74, D, 159. 123 pp.

India Meteorological Department (IMD), 1995. Rainfall probability for selected rainfall stations in India. Vols. 1 and 2, 806 pp.

Mooley, D.A. 1973. Gamma distribution probability model for Asian summer monsoon monthly rainfall. Monthly Weather Review. 10: 160-176.

Muchinda, M.R. and Venkataraman, S. 1986. Dryland farming zones of Zambia. Agrometeorological Report No. 10. Met. Dept. Lusaka, Zambia. 26 pp.

Musick. J.L.; Jones, O.R.; Stewart, B. A. and Dusek, D.A. 1994. Water yield relationship for irrigated and dryland wheat in the US Southern Plains. Agron. Jl. 86: 980-986.

Palmer, W.C. 1965. Meteorological Drought. U.S. Wea. Bur. Res. Paper no. 45, 58 pp.

Pandey, P.K.; Panda, S.N. and Pholane, L.P. 2005. Modeling for maximizing precipitation utilization in rainfed agriculture in Eastern India. Bull. Nat. Inst. Ecol. 16: 113-120.

Popov, G.F. 1984. Crop monitoring and forecasting. Proc. Int. Symp. Agrometeorology of sorghum and millet in the semi-arid Tropics. International. Crop Res, Inst. for Semi-Arid Tropics (ICRISAT), Hyderabad, India. 307-316.

Raji Reddy, R.D.; Sreenivas. G.; Narasimha Rao, S.B.S. and Yogeshwara Rao, A. 2003. Water requirement satisfaction index of groundnut under South-Telangana agro-climatic conditions. Jl. of Agrometeorology, 5: 134-137.

Ritchie, J.T. 1972. Model for predicting evaporation from a row crop with incomplete crop cover. Water Resources Research. 8: 1204-1213.

Shaw, R.H. 1974. A weighted moisture stress index for corn in Iowa. Iowa State Jl. of Res. 49: 101-104.

Shaw, R.H. 1977. Use of moisture stress index for examining climate trends and corn yields in Iowa. Iowa State Jl. of. Res. 51: 249-254.

Srivastava, R.C. 1996. Design of runoff recycling irrigation system for rice cultivation. Irrigation and Drainage Engineering, 122: 331-335.

Srivastava, R.C. 2001. Methodology for design of water harvesting system for high rainfall areas. Agric. Water Management, 47: 37-53.

Srivastava, N.N.; Victor, U.S. and Ramana Rao, B.V. 1989. Commencement of growing season and productivity of ground-nut in Rajkot district. Mausam, 40: 399-402.

Subrahmanyam, V.P.; Subba Rao, B. and Subramanian, A.R. 1965. Koppen and Thornthwaite systems of climatic classification as applied to India. Annals of Arid Zone, 4: 47-55.

Sudar, R.A.; Saxton, K.E. and Spomer, R.G. 1981. A predictive model of water stress in corn and soybean. Trans. Amer. Soc. Agric. Eng. 24: 97-102.

Thom, H.C.S. 1966. Some methods of climatological analysis. Tech. Note, 81, WMO, 53 pp.

United States Dept. of Agriculture, USDA, 1972. Hydrology. In: SCS National Engineering Handbook, Section 4. SCS Washington D.C. 142 pp.

Venkataraman, S. 1979. Agroclimatic Delineation of Dryland Sowing Periods- A case study for Karnataka. Mausam, 30: 99-104.

Venkataraman, S. 1981. Duration of Dependable Precipitation in Dryland Areas of Northwest India, Maharashtra and Karnataka. Mausam, 32: 324-325.

Venkataraman, S. 2001. A simple and rational agroclimatic method for rainfall zonations in dryland areas. Ind. Jl. Environment & Ecoplanning, 5: 135-144.

Venkataraman, S. 2003. Agrometeorological assessment of soil moisture stress for Kharif crops on weekly basis for real-time use. Jl. Cur. Sci. 2: 41-46.

Venkataraman, S. 2005. A crop-weather model for water balance studies of Kharif crops through rainfall budgeting. Ind. Jl. Environ. & Ecoplan. 10: 285-290.

Victor, U.S.; Srivastava, N.N. and Ramana Rao, B.V. 1988. Quantification of crop yields under rainfed conditions using a simple soil water balance model. Theor. Appl. Climatol. 39: 73-80.

Victor, U.S.; Srivatsava. N.N. and Ramana Rao, B.V. 1991. Application of crop water use models under rainfed conditions in seasonally arid tropics. International Jl. of Ecol. and Environmental Sciences, 17: 129-137.

Victor, U.S.; Srivastava, N.N.; Subba Rao, A.V.M. and Ramna Rao, B.V. 1996. Managing the impact of seasonal rainfall variability through response farming at a semi-arid tropical location. Current Sci. 71: 392-397.

Wickham, T.H. and Singh, V.P. 1978. Water management through wet soil. In: Soil and Rice. IRRI. Los Banos. Phillipines. 337-338 pp.

Yogeswara Rao, A.; Padmalatha, Y. and Krishna Rao, K. 1990. Quantification of groundnut pod yield in arid region of Anantapur to water availability. Jl. Res. Andhra Pradesh Agric. Univ. 18: 243-251.

Zhang, H. and Oweis, 1999. Water-yield relations and optimal irrigation scheduling of wheat in Mediterranean region. Agric. Water Management, 38: 195-211.

CHAPTER 7

Avoidance, Anticipation and Control of Pests and Diseases

Introduction

To meet the ever increasing requirements of the populace for staple food grains, protein rich protective foods like pulses and agro-industrial products like edible oils, fibres and sugar, scientists have been evolving cultivars with higher and higher unit area yields. However, the new crops and/or cultivars have been susceptible to a wide range of pests and diseases, both major and minor. The outbreaks of Red Hairy Caterpillar (Amscata Sp.), Surface Grasshopper (Oxya bidenta) and the Cotton Leaf-roller (Sylepta derogata) following the introduction respectively of groundnut, berseem and new world cottons in India are cases in point. The cultivars that ushered in the Green Revolution have been vulnerable to many pests and disease afflictions (Sardar Singh, 1968). Even genetically modified crop Cultivars are no exceptions is borne out by the following facts. The Tobacco Bud Worm in USA and the Boll Worms in India are the major pests of traditional Cotton cultivars. The Mealy Bug (Dhawan et al., 2007) and WhiteFly, Jassids and the Tobacco caterpillar (Dhawan et al., 2009) in Punjab, India and several species of Mirid bugs in China (Wang et al., 2009) have recently emerged as major pests of the GM cultivar Bt Cotton. Global crop-wise estimates of loss in potential production are equally shared between weather-induced deficits and losses due to pests and diseases. Thus, protection of crops from pests and diseases becomes an over-riding priority in crop culture. In this prophylactic measures like avoidance of pest and disease attacks should take precedence over curative ones like chemical and other measures to mitigate effects of pest and disease attacks.

As mentioned at the outset, scientific studies on pests and diseases have followed national calamities like the Potato famine due to Potato Blight in Europe and USA in the mid 19th century. Chestnut Blight in the USA towards the end of the 19th century, Rice famine in West Bengal, India in 1942 due to Helminthisporium disease of Rice and Wheat famine in Madhya Pradesh, India in 1947 due to Wheat Rusts. Developments in the study of movements in the frictional layers of the atmosphere and in the turbulent transfer processes close to the ground have contributed to an understanding of long and short distance movements of insects and disease-causing spores respectively. The role of weather in the incidence and spread of crop pests and diseases gained prominence following the successful unraveling of the long distance movements of international dimensions of (i) wheat rusts in North America, Europe and India and (ii) Desert Locust from East Africa through Arabia into the Indo-Pak subcontinent.

7.1   Dangers of Over Protection

The simple solution of applying biocides (pesticides and fungicides) as spray or dust at periodic intervals increases the cost of cultivation as the quantum of produce saved is highly disproportionate to the cost incurred. Pesticides and fungicides directly applied to the soil or entering the soil as leaf-drips and stem-flow from foliar sprays and incorporation of crop residues treated with systemic pesticides and fungicides lead to a buildup of concentrations of the chemicals in the top soil. Water seeping through or flowing over soils containing concentrated biocide residues, pollute ground water and surface water reservoirs respectively. Kler et al., (2005) have drawn attention to the accumulation of systemic biocides causing damage to the animals and humans consuming such sections of the crop. Over-protection often leads to destruction of natural enemies of pests and diseases as in case of cotton pests (Sundramurthy and chitra, 1996) and make even normally minor pests and diseases into major ones. More seriously, overuse of pesticides and fungicides leads to the development of strains of pest and disease organisms that are resistant to chemo-control. Development of new strains of larvae of Codling Moth (Cydia pomennela) developing resistance to lead arsenate sprays (De Villiers, 1966), development of strains of mosquitoes resistant to DDT and occurrence of chemo-resistant strains of wheat rusts in USA and India are cases in point. The raw material needed for manufacture of inorganic biocides is Liquefied Natural Gas, LNG, which the world is rapidly running short of. Thus, use of renewable eco-friendly organic biocides needs to be done quite soon.

7.2   Pre-Disposing Conditions for Affliction

To ensure that a crop is treated with biocides minimally and yet is assured of freedom from severe attacks from pests and diseases, it is necessary to have some advance indication regarding their likely incidence. For this a knowledge of the conditions pre-disposing a given crop for affliction by a given pest or disease is required. Weather is a main but not the only factor affecting the population dynamics of pest and diseases organisms. Crop and other pre-disposing factors that influence, in real-time, the afflictions by pests and diseases are detailed below.

7.2.1   Host Resistance

Organised attempts have been made to evolve cultivars of crops resistant to a given major pest or disease with a view to curtailing, if not eliminating, the quantum of use of pesticides and fungicides. Host resistance is itself influenced by weather. Wheat varieties which are immune to Black Rust become susceptible when day temperatures exceed 20 ℃ (Joshi, 1963). While cool soils with temperatures less than 28 ℃ make rice seedlings susceptible to Leaf Spot (Helminthosporium Oryzae), high soil temperatures make potato plants susceptible to Soft Rot (Erwinia Sp). Varieties of rice, normally resistant to Paddy Blast (Piricularia oryzae) become susceptible when night temperatures go below 20 ℃ (Sadasivan et al., 1963).

A cultivar cannot be grown solely for being resistant to a particular pest or disease. Need for higher yields and/or reduction in cost of cultivation lead to their replacement by new varieties, which are more often than not highly vulnerable to biotic setbacks. In this connection, the recent advocacy of using pest resistant Genetically Modified (GM) cultivars of crops needs a critical appraisal and the same is detailed below.

Regarding GM crops, the fears expressed by experts on risks of contamination of non-GM cultivars through escape of transgenes into natural ecosystems by pollen and seed dispersal are genuine. Despite there being no GM soybean cultivar there are already reports of traces of BT in traditional Soybean cultivars in India. Terminator technology takes care of danger through seed dispersal. However, pollen of GM cultivars has been found at distances several times greater than the prescribed barren zones. Risks are greater for crops with cross compatibility with their wild ancestors and for crops with greater intra-special genetic diversity. For example, intra-special diversity is much pronounced for Cotton in India than in China. Thus, the adverse ecological effects of Bt Cotton will be more severe in India. Many countries have or will soon have legislation banning import of GM crop produce. Thus, trade interest in export of agricultural produce remains in GM free cropping.

None of the GM cultivars have been shown to out-yield their non-GM ones. What mars GM technology the most is the inability of the farmers to reproduce the seeds, retaining their original genuineness for sowing their next crop. Thus, they are at the mercy of seed companies and are easy targets for touts pandering non-genuine F 2 material. If the cost of procurement of seeds from seed companies due to the terminator technology is taken into account, the net income from cultivation of GM cultivars will be much less than those involving normal cultivars.

Therefore, we need examine only the two main reasons advocated for use of GM crops, namely their resistance to weeds and pests. Because of climatic influences under global warming, the distribution of weed flora across regions and their composition in a region are likely to undergo significant changes and hence vary greatly. There is no broad-spectrum weedicide effective against all weeds. The indication is that over-use of weedicides for a long period of time will lead to chemo-resistance in weeds (Nandal et al, 1999). It is, therefore, difficult to visualise a transgenetically induced plant-toxin that will be effective against all weeds. So chemicals to control other weeds have to be used. Thus, a GM cultivar can be effective against one weed and at best a small group of weeds. Such weeds will soon develop a resistance to or produce mutants resistant to the GM crop toxin. The successful adoption of the System of Rice Intensification (SRI), in which the fields are kept weed-free, manually and/or mechanically, shows that GM cultivars are not needed for weed suppression.

The arguments advanced against use of GM cultivars for weed suppression apply per se to use of pest resistant GM cultivars. The above arguments will also apply if in future some company comes up with claims of a disease resistant GM cultivar. Further, the likely medium-term effects of the use of pest and weedicide resistant GM cultivars on soil environment appear to have not been studied.

Considering that substantial benefits vis-a-vis the non GM cultivars has not been established and that the fears expressed of environmental and other risks involved in GM cropping are genuine, the advocacy of GM crops for control of weeds is unnecessary and of pests is premature.

7.2.2   Food Supply

Years of good harvests are observed to be years of maximal incidence of pests and diseases. A given pest or disease sometimes has a favorite crop to feed on and multiply rapidly. Dubey and Venkataraman (1977) in a study on the likely areas of colonisation, due to accidental entry, of the Japanese beetle (Popillia japonica Newman), point out that some of the climatically suitable areas in India would be relatively free of the beetle because of non-availability of their favorite feeding material.

7.2.3   Polyphaghy

Absence of juxtapositioning of susceptible stage of the crop and the infective stage of the pest/disease has a destabilising effect on build up of pests and diseases, However, polyphagous pests i.e., pests that can thrive on more than one crop are the least affected by the above constraint. Pests (i) like the leaf-eating caterpillar, Spodoptera litura F, which can also thrive on groundnut, cotton, pulses, castor, cauliflower and several vegetables (Singh and Jalali, 1997; Ranga Rao et al., 2008) (ii) all species of Thrips (Prasad et al., 2008; Nandagopal et al., 2008). (iii) the Pink Stem Borer of cotton which can survive on maize, sorghum. Barley and wheat (iv) the White Backed Plant Hopper which can thrive on rice. Millets and grasses (Shamim et al., 2009) and (v) the white-fly which can live on cotton, sugarcane and sesame (Gupta et al., 2009) are polyphagous in nature.

Large scale cultivation of susceptible and/or the same type of crop varieties under relay cropping might have whittled down the barrier to polyphagy (Venkataraman and Krishnan, 1992) by providing prolonged availability of favourite crops and crop stages. The intensification of the Red Rot (Colletotrichum falcatum) and Smut (Ustilago scitaminea) diseases of sugarcane due to planting of canes in October and the stabilisation of Borer attacks on account of delayed harvesting and ratooning of the cane crop (Sardar Singh, 1968) can be traced to the increased availability of susceptible crop stages.

7.2.4   Alternate Hosts

Sometimes pest and disease organisms can get over the constraints of non-availability of susceptible hosts and growth stages by spending part of their life cycle in or on other compatible plants or organism. For example wheat rusts in USA continue to remain infective for a long period by switching over to Barberry grass in their long distance movement. The Johnson grass (Sorghum halpense) assists (Gupta and Kulshreshta, 1957) in the survival of the Sugarcane Stalk Borer (Chilotraea auricilia) in India. Pre-monsoon growth of grasses aids in the initial and strong build up of Hispa (Dicladispa armigera), Gall Midge (Pachydiplosis oryzae) and Gundhi Bug (Leptocorisa vericonis) to devastate the rice crop in India. Where alternate hosts can definitely be located, the agronomic advice is for pre-seasonal eradication of such crops/vegetation. The raising of tomato plants in and around citrus orchards lowers the incidence of the fruit moth as they prefer the succulent tomato to the hard-rinded citrus fruit. Pests and diseases can thrive on the post-harvest stubbles left in the field. For example, White Rust (Albugo candida) overwinters as resting spore in decaying plant tissues or as a seed contaminant or in soil (Kumar and Chakravarthy, 2008) to germinate, under optimum conditions of moisture and temperature, to infect the rapeseed/mustard crop in winter. It is this aspect that governs legislative stipulations for non-retention of crop residues in fields beyond a certain date. Mealy bug can survive on cotton stalks, barks of trees and several weeds in the off season.

7.2.5   Carriers

Carriers play a role in the transmission of crop diseases, though in a less spectacular manner than in case of human and animal pathology. Examples are: (i) the Flea Beetle as a carrier of the Bacterial Wilt of corn (ii) the White Fly as the vector for the leaf-curl virus of tomato and cotton (iii) the Aphids as the vector for many virus diseases of potato, tobacco, broad bean etc., and (iii) Thrips, as carriers of viruses causing Leaf-curl disease and Bud Necrosis of Groundnut. An interesting case of the same carrier harboring more than one disease organism with differential infective rhythms is provided by the Green Lesf Hopper (Nephotettix Sp) for the Tungro and Yellow Dwarf viruses of rice (Mishra et al., 1973.)

The Bionomics of vectors, carriers, parasites and predators of pest and disease organisms is weather controlled. Thus, the main advantage in locating carriers/vectors lies in the opportunity afforded for (i) forecasting the likely intensity of attacks through monitoring the activity level of vectors/carriers (Miller, 1959) and (ii) adjustment of sowing dates and use of varieties of appropriate durations to avoid peak periods of activity of the vectors/carriers (Sardar Singh, 1968).

7.2.6   Natural Enemies

Many insect pests and diseases are kept under check by some natural enemies which prey or parasitise on them. In this, human interference can either be beneficial or harmful. The destruction of natural enemies of pests and diseases as a corollary of over-protection of crops by chemicals has been mentioned earlier. Biological control of pests, through the introduction, from exotic regions, of their natural parasites/predators is now an accepted and widely used non-chemical method in the fight against crop pests and constitutes beneficial human interference. Biological control of diseases has not been successful in case of disease organisms, which have far lesser natural enemies compared to pests. The natural enemies are very often host-specific. The usefulness of a sound climatic approach for biological control of the Cotton Boll Weevil in U.S.A. (Pierce et al., 1912) and the neglect of the climatic aspect leading to the failure of the attempt in biological control of the Sugarcane Mealy Bug in Egypt and Palestine (Uvarov, 1931) have been mentioned earlier. In Punjab, India, planting of maize crop one to two months ahead of the normal planting time of August exposes the crop to severe incidence of Stem Borer (Chilo zonelus) since its natural enemies reach peak activity in or after August (Sardar Singh, 1968). Inter-cropping of maize with cotton had greater number of natural enemies of cotton pests and diseases than cotton alone (Patel and Yadav, 2004). However Coccinelid and Syrphid predators failed to repress Mustard Aphid population in West Bengal (Khan et al., 2008).

7.2.7 Survival

Weather factors exert a direct influence on incidence of pests and diseases. The range of weather conditions a pest or disease can endure and the congenial weather needs for quick development and multiplication vary with the infective organisms. For example the temperature conditions in Peninsular India are fungicidal to the development of Yellow Rust (puccinia striiformis) of wheat and sustained development of Blister Blight (Exobasidium sp.) of Tea. In India, the activity of Sugarcane Shoot Borer (Chilotraea infuscatellus) is arrested with the arrival of monsoon in the spring planted sugarcane tract while its activity continues in the arid irrigated areas till the onset of winter (Kalra and Sharma, 1963).

To overcome harmful weather conditions a pest or disease organism can migrate to more favourable habitats. In India, the Mustard Aphid (Lipaphis erysimi, K.) migrates from plains to hills during summer (Roy and Baral, 2004). White Rust, Albugo candida, overwinters as resting spores in decaying plant tissues or as a seed contaminant or in soil and moisture and temperature play a key role in germination of spores and in incidence and spread of the disease (Kuamar and Chakravarthy, 2008). They may also go into a state of rest with attendant physio-morphological changes. Fully grown larvae of Helicoverpa armigera, which cannot survive on the plant for more than a week, drop down into the ground and burrow to pupate a depth of 3 to 7 cm, where soil temperatures influence the duration of the pupal stage and population level, which is maximal with soil temperatures of 22 °C at 5 cm depth and 24 °C at 15 cm depth (Metange et al. 1984). The success of survival by resting depends on the severity of weather. For example, biological setbacks occurring in the cool season in the tropics are not carried forward in Autumn due to high Summer temperatures. Similarly Autumn infections are not carried forward to Spring in regions of cold winters.

7.2.8 Exotic Origin

Difficulties in over-summering and over-wintering results in the need for origination of infective stages of pests and inocula of diseases from favoured and specified locations outside the region/country. The above feature in turn calls for initiation of prophylactic measures in exotic areas for the control of such pests and diseases.

In the case of diseases, normally most of the spores fall in or near the crop (Aylor, 1986). However, some spores escape into the air due to turbulence or rain-splash. Fragmented and open crop stands release twice as much spore compared to extensive mono-cropped areas with closed crop canopies. The released spores can move long distances through the atmosphere (Gregory, 1973) before deposition by gravity, convergence or rain-washout occurs. Temperatures and moisture by way of dew, rain or high humidity determine germination of spores. However, as discussed in some detail below, crop factors have a vital bearing on disease incidence. Thus, live spores can occur without disease incidence, as reported by Krupa et al., (2006), in case of Soybean Rust in the early to mid-season soybean crop period in Northern areas of the crop belt in USA.

If the origin of spores is at a high elevation like hills, the upper air currents will carry the spores and deposit them by convergence at long distances from their origin. For example, the Uredeospores of Black Rust of wheat released in November in the Palani and Nilgiris hills of South India get carried away by upper air currents to be deposited in regions of wind convergence in the North Indian Plains (Nagarajan and Singh 1973). In such cases the areas between the points of origin and deposition of the spores will remain disease free. The primary inoculum for infecting wheat in India originates from the sub-mountainous regions of Nepal (Mehta, 1940; 1952). A comprehensive review relating to the release, movement, deposition and infection through fresh spore production relating to Soybean Rust caused by Phakopsora pachyrhizi has been presented by Emerson and Esker (2008). Mention about origin of infecting desert locusts in East Africa has been made earlier. Control of the Desert Locust involves international coordination and collaboration.

7.2.9 Susceptibility

What matters most in incidence of pests and diseases are the facts that (i) a given crop is susceptible to a given pest or disease at a particular age or stage of its development and (ii) that a given pest or diseases organism becomes highly destructive only during certain stages of its development. Some examples of the above types of situations are mentioned below.

7.2.9.1 Crop Age

Mature plants of rapeseed-mustard are not conducive for survival of the Mustard Aphid (Lipaphis erysimi K) (Roy and Baral, 2004). While Rice Hispa (Dicladispa armigera) prefers young plants that are 1 to 2 months old, only mature cotton plants are affected by white and dry root rot diseases. Young rice plants are more susceptible to Helminthosporium and Paddy Blast and resistance increases with maturity of plants (Padmanbhan and Ganguly, 1954). Two to three week old Sorghum plants are most susceptible (Dubey and Yadav, 1980; Ogwaro and Kokwaro, 1981) to shoot fly (Atherigona soccata, Rondani) whose oviposition is strongly influenced by the age of the plant (Padmaja et al., 2005). Incidence of early blight (Alternria solani (E and M) occurred at 32 plus or minus 4 days of planting in all 3 dates of sowing and all three years at Anand, Gujarat, India (Patel et al., 2004) while incidence of late blight (Phytophthora infestans) occurred from 23 to 27 days of sowing of potato in all five dates of sowing in the two years of study at Kalyani, West Bengal, India (Saha et al., 2008).

7.2.9.2 Crop Stage

Tuberization stage of potato is the most favourable (Patel at al. 2004) for initiation of the Early Blight disease (Alternaria solani, E and M). Flowering (Sharma and Kashyap.1998) and Pod Initiation (Singh et al 1984) of Brassica crops are the most vulnerable and sensitive stages respectively for incidence of the Mustard Aphid (Lipaphis erysimi, Kaltenbach). Tillering and heading stages of rice are the most susceptible (Yella Reddy et al., 2006) to attacks of Paddy Blast (Pyricularia grisea Sacc). Seedling and flowering stages of groundnut are most susceptible (Nandagopal et. al., 2006) to the leaf-eating caterpillar (Spodoptera litura F). Mature plants of potato are more susceptible (Fry and Apple. 1986) to Potato Blight (Phytophthora infestans, Montangne). Flowering stage of rapeseed mustard is the most susceptible stage for infection by white rust (Saharan et al., 1984). Late sown rapeseed and mustard suffer more from white rust (Albugo candida) than early or normal sown crops (Kumar et al. 1995) due to availability of more susceptible foliage for a longer duration in a favourable micro-environment (Kumar and Chakravarthy, 2008). Soybean is most vulnerable to rust (P. pachyrthizi) in the reproductive phase (Quales and Young, 2006) from flowering to full seed and pre-flowering spraying of fungicides is uneconomical.

7.2.9.3 Infectivity of Pests and Diseases

The first and fourth in star stages of the Lucerne Weevil, Hyper postica (Gyllenhal) cause maximum damage to lucerne (Singh et al. 2009 a). Secondary conidia constitute the infective stage of Tilletia indica causing the disease Karnal bunt of wheat by infecting the developing ovaries and hence the crop stage from awn emergence to post-anthesis becomes a vulnerable crop period for Karnal Bunt (Kaur et al., 2006). Generally pests cause maximum damage in their larval or adult stage while diseases become menacing when spore production, spore germination and mycelial growth occur continuously. Interestingly, both the nymphs and adults of the White Backed Plant Hopper and White fly cause damage (Gupta et al., 2009; Shamim et al., 2009).

Weather has a major role in the phenological development of crops. Thus, many regional peculiarities relating to absence or virulence of pests and diseases on a given crop, endemicity of areas and periods of specific pests and diseases are explainable on the basis of weather influences on the coincidence or non-coincidence of available food supply, susceptible host phase and infective organism stage.

7.3 Avoidance of Pests and Diseases

7.3.1 Identification of Endemic Areas and/or Periods

The laboratory or empirical approach, as detailed below, can help in determining the type of relationship between weather elements and a given pest or disease. Bourke (1955) advocates plotting the monthly march of parameters X and Y, that limit the incidence of a given pest or disease, in different regions to either identify areas and periods free from or endemic to them or gauge level of incidence of attack by them.   For example, in USA Soybean Rust (P. pachyrthizi) is endemic in South Eastern states but only seasonal in the Northern ones due to cold stresses during winter at the higher latitudes (Pivona and yang,(2004). However, in Brazil the climate is congenial for the soybean rust to become endemic (Emerson and Esker (2008). In India, Late Blight of Potato (Phytophthora infestans) is endemic in Punjab and Assam Hills. Delineation of the likely areas and periods of incidence of the Harvester Termite (Hodotermes mossambicus) in

South Africa (Nel, 1965), Locust in Australia (De Villiers,1966) and Cercospora leaf spot of groundnut in India (Venktaraman and Kazi, 1979) are cases in point.

7.3.2   Sowing Dates

The requirement of juxtapositioning of the susceptible crop stage and infective phase of a pest or disease organism for incidence of a pest or disease has been successfully used in avoiding severe incidence of pests and diseases by the simple strategy of change in sowing dates and/or change in life-duration of varieties. For example, potato sown around end September in North Indian Plains escape attack of Potato Blight (Arora et al., 1999) caused by Phytophthora Infestans (Montangne). Similarly since September is the most favoured month for development of Pyricularia oryzae (rice blast) and as only young plants are the most susceptible, Padmanabhan and Ganguly (1954) advocate adjustments in sowing dates to ensure that in September the post transplanting age of the plant is sufficiently advanced to avoid blast. The need for selecting the second best date for sowing to avoid a pest or disease in climatic zonation of crops has been mentioned earlier. For the same reasons, wrong choice of sowing dates and/or life-duration of cultivars can bring about an enforced coincidence of the infective stage of a pest or disease and the susceptible crop periods. For the same reason, the shortened life-duration of improved crop varieties exerts a deleterious effect on traditional pests and diseases and hence is beneficial (Sardar Singh, 1968). In Punjab, India, early sowing in October of rapeseed-mustard greatly reduced the incidence of Aphids (Dhaliwal and Hundal, 2004). Sowing of rapeseed mustard (Brassica juncea) before 15 October (Kumar and Chakravarthy, 2008) and of Potato before October (Arora et al.; 1999) in North West India has been advocated to avoid white rust and late blight respectively.

7.3.3   Mixed Crop Stands

Mixed cropping can substantially reduce intensity of incidence of pests and diseases by denying the organisms the required micro-climate associated with mono cropping and providing the one required by their natural enemies. As mentioned earlier, inter-cropping of cotton has more natural enemies of cotton pests than cotton alone (Patel and Yadav, 1996).

7.3.4   Upwind Sowing

In staggered sowings, planting of later crops upwind of the earlier ones helps avoid the build up of pests and diseases from earlier sown crops affecting the later ones.

7.4   Anticipation of Outbreaks of Pests and Diseases

Agronomic measures adopted to avoid pests and diseases are not necessarily the best from the point of view of realising the yield potential for the cultivar at that place and season. A farmer would like to go in for a strategy that ensures that the increase in yields far outweigh (i) the cost of control operations against pests and diseases rather than the decrease in yields due to safe sowing dates, mixed cropping etc. For this, control measures must be minimal but effective. Since both crop and weather factors, directly and indirectly, influence afflictions by pests and diseases, the need for and usefulness of advice for control operations, only when present crop cum anticipated weather conditions warrant it, needs no special emphasis.

7.4.1   Approaches

The methods used to anticipate pest and disease outbreaks of crops can broadly be grouped under two heads, namely non-weather based and weather based. Amongst the former are the geographical method, phenological technique and biological indicators while the weather-based methods include fundamental laboratory studies and the empirical field approach (Bourke, 1955). Details relating the above are given below.

7.4.1.1 Geographical Method

This method is mainly used for disease-anticipation. In this, the areal progression of the belts of affliction is used to anticipate disease appearance in geographically contiguous areas when infection is noted in one part of the region. Examples are: (i) occurrence of Potato Blight in North Japan following its appearance in South Japan (Salmon,1951), (ii) the southerly progress of Paddy Blast along the eastern coast of India (Padmanabhan,1965) and (iii) northerly progress of Black Rust of wheat in India from its appearance in the South (Joshi and Palmer, 1973). In U.S.A. Rusts of Soybean, Corn and Wheat spread from south to north in the growing season and incidence of rusts in the Gulf States (Alabama, Florida, Lousiana,

Missisipi and Texas) and Georgia can be used to assess progressive risk of disease incidence in the northern states and make informed decision on use of fungicides.

7.4.1.2 Phenological Technique

In this the appearance of a biological phenomenon is used to anticipate a pest or disease outbreak. Examples are the relations between (i) the first date of blooming of Cherries to incidence of Rice Stem Borer in Japan (ii) the infection of the Sycamore tree by the Anthracnose fungus (Gnomia veneta) to Grapevine Mildew in France (Darpoux, 1943) and epiphytic infestation of the Cacao tree (Theobroma Cacao) to the Black Pod disease (Phytophthora palmivora) in Nigeria (Thorold, 1952).

7.4.1.3 Biological Indicators

Determination of bacterial population in the waters of paddy fields for anticipating the occurrence of Bacterial Blight disease of Paddy (Xanthomonas oryzae) have been suggested by Yoshimura and Tagami (1967). In India, Rao and Srivastava (1973) found that phage population in rice fields could be used for forecasting the Bacterial Blight disease.

7.5   Laboratory Studies - Usefulness and Limitations

Organisms which cannot maintain their body temperature on their own are called Poikilothermal. Such organisms are highly affected by the temperatures to which they are exposed. Insects and to a lesser extent disease organisms are Poikilothermal. The above fact has facilitated laboratory studies for the determination of (i) the maximum, minimum and optimum temperatures for development for various growth stages and (ii) heat unit requirements above a base temperatures for the completion of specified growth phases for many a pest or disease organism.

7.5.1   Biographs

The preparation of Biographs from laboratory data to take account of the differences in rates of development of insects at different temperature levels has been advocated by Pradhan (1946). The Biograph is essentially a thermogram in which the temperature graduations are replaced by lines of average velocity of development corresponding to the temperature. The Biographs can either cover the full life cycle of an insect or may be prepared for a specific developmental stage of the organism. The manner of preparation and use of the Biograph have been detailed by Dang and Doharey (1971). The Biograph can assist in determining the rapidity of completion of the full life-cycle of a given insect from any starting point and hence help ascertain the possible number of generations in a given period. The idea of Biograph is also applicable to disease causing organisms.

7.5.2   Accumulated Degree-Days (ADDs)

Extensive laboratory studies have been carried out to study the suitability of the application of the concept of accumulated degree days to pests and disease organisms for their phenological development from one stage to another. The dependency on degree-day accumulations of the development of the ontogenic phases of several insects and plant pathogens have been reported.

The threshold maximum and minimum temperatures and ADDs for growth stages of many pests and diseases, determined in the laboratory, show that these not only vary amongst organisms but also for different developmental stages of the same organism. The variability in base temperatures with growth stages as and when present may at first sight appear daunting. However, if the variations in the main and long growth stages are not far different, assumption of a mean base temperature will not affect much the computed duration in days of growth stages; For example, egg and maturity stages are very short compared to the nymph and adult stages of insects. So as long as the base temperatures for the latter two stages are not too much different, use of a base temperature as mean of the nymph and adult stages will suffice in practice.

7.5.2.1 Diurnal Variations of Temperature

Laboratory studies to determine threshold temperatures and ADDs are carried out at constant temperatures. However, in the field, the temperatures show a diurnal variation. The later feature leads to lower ADDs than laboratory determined values. For example for Codling Moth (Cydia pomennella) the field values of ADD for the stages from Egg to Larva and Larva to Pupa were seen to be lower than the laboratory values (Howell and Niven, 2000). However, Ranga Rao et al., (2008) report that the response of various stages of

Spodopteras litura to temperature under field conditions was similar to that under constant laboratory conditions. The above are due to the level of temperatures in the field. If the temperatures are in the range of linear relationship of ADD with temperatures, field and laboratory values will agree. However, if temperatures in the field lie in the curvilinear section of the temperature versus ADD curve, at the higher or lower levels, the field and laboratory values will not agree. Again, the nature of relationship between phenological development of a pest or disease may vary with growth stage. For example, Podosphaera pannosa causing Powdery Mildew of Roses has a non-linear relationship with temperature in the incubation period and a linear one in the post-incubation period (Xu, 1996).

7.5.2.2 The crop factor

However, in applying laboratory data for field use, the extent of modification of the macro-climate by crop cover, as mentioned in quite some detail in an earlier chapter, must be allowed for as has been done by Broadbent (1950) and Waggoner (1960) in case of Potato crop. Leaf wetness duration is the main factor affecting disease incidence. Because of greater nocturnal cooling of crop foliage, actual duration of leaf wetness will be significantly higher compared to those based on screen minimum. Del Ponte et al., (2006) found severe epidemics of soybean rust to occur in regions of Brazil where day time air temperatures were higher than the upper limits found under controlled conditions. In situations when the crop stand is open and in rainy season, when the crop micro-climate is close to the macro-climate, laboratory studies can straightaway be put to field applications.

7.6   Field Studies

To help specify the pre-disposing weather conditions for incidence, multiplication and spread of diseases, it is first of all necessary to account for the effects on non-weather factors for the presence or absence of pests and diseases, The influence of biotic factors have been elucidated in some detail to highlight the daunting nature of field studies to segregate the effects of and ascertain the relative influence of biotic and abiotic factors on population build up of insects and disease-severity factors so as to ascertain the weather relations of pests and diseases. For example, Nandagopal et al., (2008) in case of population of Sesbania Thrips (Caliothrips Bagnall) in Saurashtra, India and Gedia et al., (2008) in case of egg masses of Spodoptera litura, report that weather factors accounted only for 47% of variations of population and egg masses. Samui et al., (2008) found that weather factors accounted for only 53% of variations in the first generation population levels of the Rice Stem Bor.

7.6.1   Limitations in Availability and use of Data

For crop-weather relationship studies, phenological and phenometric data on crops can be routinely obtained from sowing date trials conducted at a net work of stations. Such data from stations located in the same crop-climate zone can be pooled for analyses so as to arrive at findings in a short period of time, say 5 Years. As in the case of study of weather-relations of growth, development and yields of crops, an examination of data on intensity and extent of incidence of pest and disease afflictions with concurrent meteorological data is necessary. Such data must relate to an unprotected crop. So pest and disease data are obtained from untreated control plots in experiments on the efficacy of chemical and/or agronomic measures in controlling incidence of pests and diseases. The direct and indirect influences of weather on and importance of juxtapositioning of infective stage of a pest or disease and the susceptible crop stage for incidence of pests and diseases had been mentioned earlier. Weather predominantly determines life-duration and phenological development of crops. The temporal march of weather parameters over short time-units, like a week, show wide variations (i) across locations in a season and (ii) between years at a place. Such weather vagaries often disturb the (i) juxtapositioning of susceptible crop stages and infective stage of pest and disease organisms (ii) checking influences of natural enemies of pests and diseases and (iii) the survival chances of pests and diseases on alternate hosts. For any given crop pest or disease, the above result in sporadic attacks across (i) locations in an area and (ii) across years at a location. Thus, most of the studies on weather aspects of incidence of pests and diseases are per force single station studies.

To maximise availability of data for analyses, recourse is taken to observe incidence of pests and diseases sown on two or three days. This technique is available for irrigated crops only. However, unlike in crop-weather relationship studies, pooling of data from different sowing dates poses some problems where the incidence of a pest or disease is governed by crop age. For example incidence of Late Blight of Potato at Kalyani. West Bengal occurred around age of 25 days after planting which ranged form 11 November to 5 December (Saha et al., 2008). On 3 sowing dates, 2 varieties and 3 years, first initiation of Early Blight of Potato occurred after 30 days of planting (Patel et., al. 2004). Age of the plant influences the incidence of shoot fly on Sorghum (Padmaja et., al , 2005). Incidence of white Rust on Mustard occurred on the same day for 2 varieties and 2 sowing dates and the time taken to attain maximum intensity was 8 weeks in all cases (Kumar and chakravarthy, 2008). Pooling of data from other stations can lead to misleading results if the value of any given weather parameter is above a threshold in one place and below in another place. For example, a positive correlation between Helicoverpa armigera population and bright sunshine hours was reported for Mid Gujarat area (Chaudhri et., al. 1999) while in Punjab a negative correlation was found between bright sunshine hours and Helicoverpa armigera population (Dhaliwal et al, 2004). The same applies to pooling of data on pests and diseases on crops like rice, groundnut and maize grown in different seasons at a place. For example, the incidence of Sesbania thrips on Groundnut in Junagadh, Gujarat, India was (a) least in the rainy season due to washout action of rainfall (b) high in Summer because of higher temperatures and low relative humidity and (c) moderate in winter due to   less rain (Prasad et al, 2008).

7.6.2   Insect and Spore Trap Catches

The realisation that occurrence of congenial weather conditions can lead to epidemic outbreaks only if there is an initial insect population or incipient disease inoculum to take advantage of the favourable weather situations, has led to the organisation of surveys for early detection of the presence of a pest or disease in the infective stage. Visual surveys can cover only limited areas in a given time period. Surveys for detection in infective stages of specific pests and diseases on a micro-scale and real-time basis will be prohibitively costly. To avoid the inherent errors in sampling in such surveys and to obtain data representative of small areas, entomologists and mycologists have been advocating the use of devices to trap insects and spores. Light traps, which are mechanical devices consisting of high wattage bulbs and suitable pest-trapping arrangements fitted to an iron mast are exposed over cropped fields at a suitable height. In later years pheromone traps (Ranga Rao et al., 1991) consisting of a plastic funnel treated with a chemical to attract male moths (Tamaki et. al., 1973) with a polythene sleeve with a chemically treated cotton plug to kill the trapped moths have come to be used. The type and efficacy of spore traps have been discussed by Sutton and Jones (1976) and Wili (1985).

Data of insect and spore traps have been used to study the effect of weather factors on pest population and disease spore fluctuations (Samui et al., 2004; Nandihalli et al., 1989; Chaudhari et. al., 1999). Significant and positive correlations between light trap catches and field populations of the Brown Plant Hopper, BPH, Nilaparvata lugens(Stal.) (Jeyrani et al., 2000), and nymphs and total macropetrous males and macropetrous total adults of BPH (Krishniah et al., 2006) have been reported.

Singh and Sachan (1993) found pheromone trap catches of male moths effectively monitored population fluctuations of Spodoptera litura F. Thus, catches in pheromone and light traps can be used to forewarn attacks by pests. The role and usefulness of trap-catches of insects and spores in forewarning of their incidence is discussed subsequently.

7.7   Forewarning of Pests and Diseases

7.7.1   Pest/Disease Weather Diagrams

Historical and concurrent data on pests and weather have been used to prepare Pest-Weather Calendars (PWCs). (Samui et al., 2004). In the PWCs the top portion highlights the weather requirements of the pest for its development. The middle portion details the week by week normal features of weather at the station. The bottom portion shows the months and weeks along with the normal times and duration of phenological phases for the crop along with the life history and times and duration of important pest stages. Historical data can similarly be used to prepare Disease–Weather Calendars (DWCs).

The use of the Standard Meteorological Week (SMW) as a time unit in preparation of PWC is not logical as incidence of pests are influenced by crop age and stage on the one hand and infective-stage phenology of the pest on the other. Because of weather vagaries, even under irrigation, the actual crop-weather-pest situation position in the standard weeks in any year will be different from that depicted in the

PWCs. The deviations will be more marked under rainfed conditions. This is akin to the rightful contention that the temporal march of weather on a short-time unit as represented by the climatic average is the one that is never realised. Thus, PWCs or PDCs have no intrinsic merit in forewarning of pests on even on a local scale and are not of much help in ascertaining for any given year the type and extent of deviations from the normal of the weather, crop and pest situation. In view of the changing nature of crop biota, use of crop information more than a decade is not advisable. It would be more useful if, Pest/Disease weather diagrams are prepared every year for at least a decade of the recent past to portray the situation in terms of crop age and crop stages right from date of sowing to harvest. Such a compilation will help in an assessment of the impact of weather and the pest/disease situation on crop prospects at the end of the crop period.

7.7.2   Local Studies

An intelligent examination of pest/disease free years/sowing dates vis-a-vis years/sowing dates of heavy incidence can throw up much useful information. For example, the Humidity/Temperature ratio was seen to be the operative factor in the incidence of Karnal Bunt of Wheat (Mavi, 1992; Jorhar et al., 1992; Dhaliwal et al., 2004).

In this either single station studies in which the emphasis is on assessment of meteorological conditions conducive to incidence or absence of a pest or disease or multi-location studies where the emphasis is on delineation of weather conditions leading to changes in periods and/or intensity of infection may be used. The work on Piriculatia oryzae in Orissa, India (Padmanabhan, 1965) comes under the first category. A multi-location study is to be preferred to a single station study. The work in India on (i) Cercospora leaf spot of groundnut in Maharashtra (Sulaiman and Agashe, 1965 (ii) on sugarcane shoot borer in Uttar Pradesh and Rajasthan (Kalra and Sharma, 1963) and (iii) on Sorghum shoot fly in Karnataka (Usman, 1968), Tamil nadu (Ponnaiya, 1951; Subbiah and Mohamed Ibrahim, 1968), Delhi (Jotwani et al., 1970), Rajasthan (Kundu et al., 1971) and Maharashtra (Raodeo and Muqum, 1971) belong to the second category.

However, great care is required to be exercised in transposing empirically derived warning criteria from one region to another. For example, the warning criteria for late blight of potato differ in the

European countries (De villiers, 1966). A similar situation would obtain in warning for potato blights in plains compared to hilly terrain. For late blight of Potato the emphasis is on (a) 10 hour periods of R.H. Greater than 90% in North and South America (Wallin and Hoyman, 1954) and (b) on occurrence of 2 days with afternoon R.H of more than 70% in South Africa (Young 1966).

7.7.3   Thumb Rules

Locally, analyses of data on pests and diseases during year(s) and/or dates of sowing of light to nil incidences with those of heavy to severe incidence have led to some useful location-specific formulation of thumb rules for anticipation of their incidence. Some such thumb rules are enumerated below.

Rice Blast, Piricularia Oryzae

At Cuttak, Orissa, India Padmanabhan et al., (1971) advisee spraying on occurrence for a week of minimum temperatures below 24 °C cum morning relative humidity of 90% or more.

Mustard Aphid. Lipaphis erysimi (K)

Chakravarthy and Gautam (2004) have proposed day degrees accumulations above a base temperature of 5 °C from 1 January for fore-gauging from 25th January the Mustard Aphid infestation level three weeks ahead at New Delhi.

Cumulative Degree days above 5 °C

Date	Degree Days (Severity)	
15 January	90 (High)	140 (Low)
20 January	115 (High)	190 (Low)
25 January	150 (High)	245 (Low)

Brown Plant Hopper (BPH) Nilparvata lugens (Stal)

Varma et al., (2008) have used cumulative rainfall in August for advanced assessment of spore trap populations of BPH in the Krishna and Godavari districts of Andhra Pradesh, India as per the following criteria:

Cumulative Rainfall in August	BPH Population Level in September
Less than 100 mm	Low
100 to 200 mm	Moderate
200 to 300 mm	High
More than 300 mm	Extremely High

Powdery Mildew of Grapes

For late pruned Grape crop in Northern Karnataka, India, the following formula has been advocated by Jahgirdar et al., (2001) to forewarn of incidence of Powdery mildew two weeks in advance, namely: DI = 86.69+ 6.41 Max T(2) +0.30 RF(4) , Where DI is Disease Index, Max T(2) and RF(4) are maximum temperature two weeks before and rainfall four weeks before disease incidence.

Pod Borer, Helicoverpa Armigera

Using rainfall deficits or excess in the period June to September, A and October and November rainfall, B respectively in Karnataka and Andhra Pradesh the following criteria have been suggested (Das et al., 2001; Trivedi et al., 2005; Das, 2013) for predicting Pod damage due to Helicoverpa armigera in Karnataka and Andhra Pradesh, India.

Rainfall Situation	Level of Pod Damage
A surplus plus B Deficit	Low
A Surplus plus B Surplus	Moderate
A Deficit plus B Deficit	Moderate
A Deficit plus B Surplus	Severe

White Rust, Albugo Candida on Brassica

Kumar and Chakravarthy (2008) have formulated the following thumb rule for appearance of White Rust at New Delhi, India.

If on 10 consecutive days there is a concomitant occurrence of (a) 150 hours of temperature in the range 10 to 20 °C (b) 180 hours of relative humidity greater than 80% and (c) less than 10 hours of bright Sunshine disease would appear.

If there are rainy days in December and January and total hours of bright sunshine is less than 40 in a 20 day period, disease would appear. As sunshine is highly correlated to disease incidence and as the other two parameters are not ready to obtain they suggest that use of sunshine criteria will suffice.

7.7.4   Multiple Regression Approach

Initial studies in weather aspects of pest and diseases consisted in using historical, concurrent data of weather and pest/disease to evaluate correlations between various weather parameters and pest population/disease severity and fitting regression equations of weather parameters that explained maximum variations in pest population/disease severity. The basic assumption behind such an approach is that the data sets of weather, pest or disease formed a composite population and insertion of realised weather data of later years would give a measure of pest population/disease severity of that year at that location. However, even on a location-specific basis, such regressions fail because of high variability of weather from year to year on a short-time unit basis, absence of initial inoculum or incipient pest population in the field at the assumed time and non-inclusion of parameters relating to the phenology of the crops, pests and diseases.

7.7.5   Regressions with Lead Time

Most importantly such multiple regressions have no forecast values. For that we need regressions with lead-time in which a pest or disease incidence will be related to the weather conditions in the preceding weeks. Some regressions with provision for lead time for predicting some rice pests for early warning are presented below.

Samui et al., (2004) give the following equation for predicting attacks of rice Pests at Pattambi, Kerala, India.

Gall Midge (Orseolia oryzae)

$$P = 726.6 - 20.8 \, X_1 - 2.2 \, X_4 \, 0 \, 16.7 \, X_5$$

where P = Gall Midge population at 38th Standard Meteorological Week, SMW; X_1 = T Max for 36th SMW, X_4 =Minimum Relative Humidity for 36th SMW and X_5 = Bright Sunshine Hours for 36th SMW

Leaf Folder (Cnaphalocrocis medinalis) at Pattambi, Kerala, India

$$P = 146.54 + 14.71X_1 - 7.9 \, X_2 - 2.03 \, X_4$$

where P = Leaf Folder Population at 48th week; X_1 = T max for 46th SMW; X_2 = T min for 46th SMW and X_4 = Minimum Relative Humidity for 44th SMW

7.7.6   Dynamic Cumulative Weather Based Index (DCWBI)

Yella Reddy et al., (2006) developed a novel approach for assigning weights based on the level and relative importance of the meteorological parameters affecting intensity of infection of rice Blast caused by Pyricularia oryzae Sacc, namely maximum and minimum temperatures and relative humidity, rainfall, rainy days and bright sunshine hours. The sums of weighted values of the above parameters expressed as a fraction of the total possible weights was termed the Dynamic Cumulative Weather Based Index (DCWBI). Calculation of DCWBI is simple and can be done for different periods like 3 days, a week or 10 days. The Dynamic Weight of Blast (DWBLA) was calculated by multiplying Blast disease severity and DCWBI. They found that at Palampur and Hyderabad, India that if DCWBI was greater than 0.50 for any week during the susceptible stages of the crop, namely tillering and heading, Blast occurred during the following week. They have given an equation for predicting Blast severity one week ahead, from values of DWBLA of previous week and DCWBI of current week. The model is simple to use and can be adopted/adapted for use in all locations.

7.7.7 Inclusion of Crop Parameters

Ideally a regression with lead time should provide for crop parameters. Such a regression for predicting Egg Counts/plant of Shoot fly of Sorghum (Atherigona soccata (Rondani)) and Dead Heart percentage caused by the Sorghum Shoot fly at Hyderabad, India during the rainy, Winter and Summer seasons presented by Padmaja et al., (2005) is given below.

DH% = – 23.53 + 17.4 (Cu Egg) + 0.34 GDD – 0.36 (Cu RF) – 0.18 (Cu VPD)

Where DH% = Percent Dead hearts in the coming week; Cu Egg is the cumulative egg count at weekly intervals starting from 2nd week after germination and given by the relation:

Egg = – 5.58 + 0.046 (ERH1) – 0.39 (SSH 1) + 0.11 (DAE) where Egg is Shoot fly eggs per plant in the coming week; ERH1 is the mean afternoon relative humidity of previous week; SSH1 is the mean sunshine hours of previous week; DAE is the age of the crop in days after emergence;

GDD is Growing Degree days above a base of 12 °C from date of sowing; Cu RF is Cumulative daily rainfall in mm from date of sowing and Cu VPD is Cumulative vapour pressure deficit in KPa from date of sowing. The regression predicted 58 and 91% respectively of variations in egg density and Dead Hearts.

7.7.8   Aptness of One Week Lead Times

Several workers had reported that as lead time increases the accuracy of predictions goes down. Thus, longer lead times would call for weather forecasts covering such lead times. As the period of validity increases the inaccuracies in weather forecasts also increase. At the moment only medium range weather forecasts of 7 to 10 days' validity can be availed off for agromet advisory work. Thus, more accurate predictive regressions with one week lead time is called for. Again, many a time a careful watch of the crop-weather situation can lead to identification of such periods. For example Singh et al., (2008) found at Jhansi, India that when the Sorghum plants were one month old, maximum temperature of 31 to 35 °C, with morning humidity of 90% and afternoon humidity of more than 50% led to disease initiation of zonate leaf-spot in the next week. Rathi et al., (1999) also found average maximum of 31 °C, and mean relative humidity of 68% favored initiation of leaf spot disease in Sorghum. In case of Ascochyta blight of Chickpea in Ludhiana, Punjab, India, for the crop period was from 1 to 17 SMW, Meteorological conditions in the 4th and 5th weeks and in the 9th to 12th week period had a significant bearing on disease severity (Bal et al., 2008). Thus, the first period could be used to forewarn of appearance of disease symptoms a month in advance and the second period helped in prediction of disease appearance. Gupta et al., (2009) report that White fly on Sesame sown in early July in the Bundelkhand region of India attained a peak in the 34 to 37 SMW which was highly correlated with maximum and minimum temperatures and afternoon relative humidity in the 30-33 SMWs and claimed that their predictive equation explained 98% of the variations in White Fly populations. Gedia et al., (2008) found weekly weather to have a significant effect on the pheromone trap catches and oviposition on cotton foliage of Spodoptera litura in the succeeding week.

7.7.9   ANN Techniques

In any concurrent data series on Crop-pest/weather, the relationship of antecedent weather influences on incidence of pest/disease is often not apparent and lies sufficiently hidden to be revealed by conventionally used statistical analyses. Thus, for better analyses of data and hence better forecasting, the Artificial Neural Network (ANN) technique is recommended to be applied. While brief details relating to the methodology and philosophy of ANN are indicated by

Mehta et al., 2001, details relating to processing and transformation of data for use of ANN technique have been given by Gupta et al., (2003). Ranga Rao et al., (2008) report that for leaf-miners, caterpillars and borers, while the accuracy of regression models and Bayesian classification were 35% and 75% respectively, the ANN model predicted the pest attacks one week in advance with a correlation coefficient of 0.96. Application of ANN to forewarn various aspects of many crop pests and diseases at selected centres has been detailed by Mehta et al., (2001).

7.8   Forewarning of Pests

7.8.1   The Accumulated Degree Days (ADDs) Approach

The calculation of accumulated degree-days (heat units) and use of the same as an aid in the management of pests (Zalom et al., 1983; Wilson and Barnett, 1983) and in forecasting initiation and peak of pest attacks during the crop-life period (Roy et al., 2002) have been mooted. Use of criterion in anticipation of Mustard Aphid has been mentioned earlier. In Colorado, USA, a simple day degree model to predict appearance of Early Blight on potato caused by Alternaria solani and advice initiation of control operations has been detailed by Franc et al., (1988).

Many field studies relating to the use of ADDs in anticipation of incidence, peaks, levels of insect populations/disease severity have been carried out for many pests and diseases. In such studies commencement of computation of ADDs are often sought to be related to calendar dates. In view of the intra-year variations in crop and weather situations even at a given place, use of calendar date is not logical and runs the risk of missing out on the vital first protective spray against the concerned organism. This will be a serious lapse as pests and diseases are more susceptible to insecticides and fungicides at the egg-hatching and spore stages.

7.8.1.1 Biofix

The above problem is sought to be solved by the concept of Biofix (Riedl et al., 1976), which is a biological marker that initiates the beginning of GDD calculations. In case of insects, it is usually the date of a particular event like the flight of a moth or capture of a pest in a trap. Biofix date does not begin with the date of first appearance of the specified event and the emphasis is on sustained occurrence of the specified event, to get over the possibility of false starts due to weather vagaries early in the season.

7.8.1.2 Economic Threshold Level

For fixing the criteria of Biofix date, the concept of Economic Injury level (EIL) come into play. Pierce (1934) had raised the question of assessment of the stage at which a pest caused economic damage. The idea of Pierce (1934) led to the development in later years of the concept of Economic Injury Level (EIL) which is defined as the lowest population density at which a pest will cause economic damage and lead to crop losses being greater than the control operations. Now EIL can vary with many factors. For a given pest and crop, EIL will be low for an emerging crop compared to an established crop. For a pest affecting the valuable part of a crop, EIL will be lower than those affecting the non-economic part. The EIL populations for borers affecting pods in legumes and ear heads in grains will be lower than those affecting leaves and stem. However, in sugarcane the EIL for stem borer will be lower than for stem borers of other crops. Again, for insects like Thrips that act as vectors or carriers EIL has to be very low in view of the double damage inflicted by their feeding on the crop and transmitting viruses.

The concept of Economic Threshold Level (ETL) had been proposed by Stern et al., (1959) as the insect number (density or intensity) when control operations must be initiated to prevent the pest population to increase and reach EIL. Farrington (1977) had suggested the benefit : cost ratio as a criteria for determining ETL. Pedigo (1991) had suggested that ETL be taken as 75% of EIL. In actual practice determination of EIL for a given pest-crop combination is less complicated.

Helicoverpa armigera is a voracious and polyphagous pest of a number of agricultural and horticultural crops and hence a menace for many crops, areas and in many seasons. Studies on EIL of Helicoverpa Armigera in Bangladesh (Zahid et al., 2008) indicate that the EIL for the pest on Chickpea is 1.1 larvae per meter row. Chaudhary and Sharma (1982) in Haryana, India and Whitman et al., (1995) in South India found the ETL of Helicoverpa armigera in gram crop to be 1.0 larva per meter row. Prabhakar et al., (1998) found ETL of H. armigera in chickpea to be higher under irrigation at 1.2 larvae per meter row compared to 0.9 larvaw per meter row for a rainfed crop. The ETL for the Mustard Aphid is reported as 30 Aphids per plant on 30% of the plants (Nayban and Chowdhury, 2002), For the Tomato fruit worm, Helicoverpa zea, the ETL is stated as 4 white eggs per 30 Tomato leaves (Zalom et al., 1990). In fact the concept of ETL is accepted as a vital component of Integrated Pest Management, IPM, which seeks to combine cultural. Biological chemical and physical measures in a manner to minimise environmental and health risks in a sustainable way. The training of farmers in use of ETL in IPM programmes has been found to be hugely successful In Maharashtra, India (Kharbade et al., 2007) and Gujarat India (Vadodaria et al., 1997).

7.8.1.3 Biofix and ETL from Insect Trap Catches

It has come to be realised that the presence, density and stage of the pest species in trap catches can be utilised to anticipate the outbreak of pests and that insect trap data would be a valuable component of IPM. Monitoring of catches in the pheromone traps as a part of IPM package for control of the gram pod borer (Helicoverps armigera) in Nalgonda and Mehboobnagar districts of Andhra Pradesh, India has been advocated (Grace et. al., 2009). As weather factors largely influenced seasonal incidence of pink bollworm of cotton (Jha and Bisen, 1994) and as Pheromone trap catches have been reported to be highly correlated to the actual field infestation of the Pink Boll Worm (Pectinophora gossypiella) of cotton (Singh and Lather, 1989; Qureshi et al., 1993), the need for and value of insect trap catches is obvious. Again, Saravanan et al., (2008) observed that while the urid crop, V. Mungo, can be infected by Powdery Mildew caused by E. Polygoni under a wide range of environmental conditions, subsequent development was mainly governed by the temperature factor. In such a case early detection of the disease spore in spore traps hardly need any emphasis.

For any pest, assessment of time of attainment of ETL through field sampling is a daunting task if it is to be an ever-going process. Insect trap data can facilitate effective field sampling. For example Hoffmann et al., (1991) observed a rapid increase in egg masses of the Tomato Fruit Worm in a narrow time window after mid august. Thus, insect trap catches of adults of the Tomato Fruit Worm becomes important in determining when plant sampling should begin for foliar egg count as control measures need to be initiated when the threshold reaches an ETL of 4 white eggs per 30 Tomato leaves (Zalom et al 1990). Since the above pest is polyphagous and attacks cotton, corn and soybean, the findings assume additional significance.

Some of the interesting Biofix criteria found in the literature are:

Cotton Pink Boll Worm, Pectinophora gossypiella: 8 moths/day/ trap for 3 consecutive days

American Boll Worm, Helicoverpa armigera: 5 moths/day/trap

Codling Moth, Cydia pomenella: 5th moth in the trap

Oriental Fruit Moth: More than 1 moth per trap

7.9   Forewarning of Crop Diseases

7.9.1   Critical Disease Level (CDL)

The concept of critical level for diseases is different from that of ETL for pests. CDL relates to the stage of development of a disease where applications of fungicides before it is not required and will be wasteful, while applications after CDL will be ineffective. The CDL for leaf blight of onion caused by Botrytis squamosa is an average of one lesion for ten leaves (Shoemaker and Lorbeer, 1969). The appearance of plant lesions was preceded by two weeks by the collection of conidia of Botrytis squamosa in Hirst spore trap (Shoemaker and Lorbeer, 1977) thus providing more than two weeks time for advising onion farmers on the timing of the initial spray. The CDL for Alternaria solani, causing Early Blight of potato is the time of the secondary sporulation coinciding with the first appearance of airborne spores (Douglas and Grasshopp, (1974).

7.9.1.1 Use of Spore Trap Catches in Anticipation of CDL

The importance of spore trap data in such a circumstance is obvious. Incidence of Soybean Rust caused by Phakospora pachyrthizi is dependent on 3 conditions, namely Soybean in the reproductive phase, cool and wet weather and viable soybean rust spores (Quales and Young, 2006). Thus, spore traps helps researchers to watch for soybean rust. Harrison et al., (1965) highlight the usefulness of spore trap data as a guide for initiating control operations against Alternaria solani causing early blight in Potato in Colorado, USA.

However, catches of spores in traps are not useful if, as in case of Downey Mildew of Watermelons (Pseudoperonospora cubensis), the spores do not get trapped until the appearance of symptoms in the field (Schenck, 1968).

7.9.2   Use of Routine Temperature Data for Disease-Forewarning

Leaf Wetness Duration (LWD) is emerging as major input in many crop disease forewarning models. Data on measured LWD, as per standardized instrument and its exposure (Sentelhas et al., 2004) are quite sparse. For obtaining LWD from weather parameters, physical and empirical models have been attempted to be used. The physical models relate to use of energy balance principles to simulate deposition of water on plant surface and its evaporation thereof (Monteith and Unsworth, 1990). The empirical models involve regression trees (Gleason et al., 1994), neural network (Francl and Panigrahi, 1997) and Fuzzy logic (Kim et al. 2004). Dew Point Depression (DPD) (Gillespie et al., 1993) with additional values of DPD to allow for evaporation of Dew from plants (Rao et al., 1998) and Number of Hours of Relative humidity Greater than 90%, NHRH > 90% (Sentelhas et al., 2008) have been suggested as measures of LWD.

There are quite a number of disease warning models that call for hours of temperatures and/or Relative humidity above specified thresholds. Use such models as well as those using DPD and NHRH will require availability of hourly data of temperatures and relative humidity on a daily basis and pose a daunting task in culling out the required information from such data.

Exponential (Adalberto, 1991) sinusoidal (Parton and Logan, 1981) and empirical (Sadler et al., 1977) models have been used to work out the diurnal pattern of hourly temperatures. Srivastava et al., (2010) have used the sinusoidal model for day time hourly temperatures and the exponential model for hourly night-time temperatures. Such work relate to computation of mean weekly or mean monthly temperatures. They require location-specific calibration. Accessory data required for use of the equations are not readily available.

The only temperature data available on a ready and daily basis are the maximum, minimum and dew-point temperatures. Use of maximum and minimum temperatures to estimate the number of hours during which temperatures of a given threshold can be expected, has been enunciated by Neild (1967).

The hourly temperatures follow two different cosine curves the ascending one from sunrise to 1300 hours and a descending one from 1300 hours to sunrise of next day and solutions of equations relating the above curves have been used to get mean daytime and mean night-time temperatures (Robertson, 1983).

Venkataraman (2002) found night-time temperature to be strongly influenced by time of sunrise and has formulated a methodology for computation of hourly temperatures from maximum and minimum temperatures. The method consists in algebraically adding to the mean temperature, specified fractions of the temperature range (maximum minus minimum) to each of the 24 hours after each of the three hours of sunrise at 5, 6 and 7 hours. The formulations of Venkataraman (2002) are presented in Table 7.1.

Table 7.1 Fraction of range to be algebraically added to mean to get hourly temperatures at different sunrise hours

	Sunrise Hours				Sunrise Hours		
T				**T**			
	5	**6**	**7**		**5**	**6**	**7**
T +1	-0.46	-0.45	-0.43	T +13	0.28	0.22	0.17
T + 2	-0.35	-0.31	-0.25	T +14	0.19	0.14	0.09
T + 3	-0.19	-0.11	0	T + 15	0.1	0.05	0
T + 4	0	0.11	0.25	T + 16	0	-0.05	-0.09
T + 5	0.19	0.31	+0, 43	T + 17	-0.1	-0.14	-0.17
T + 6	0.35	0.45	0.5	T + 18	-0.19	-0.22	-0.25
T + 7	0.46	0.5	0.5	T + 19	-0.28	-0.3	-0.32
T +8	0.5	0.49	0.47	T + 20	-0.35	-0.37	-0.38
T + 9	0.49	0.47	0.43	T + 21	-0.41	-0.43	-0.43
T + 10	0.46	0.43	0.38	T + 22	-0.46	-0.47	-0.47
T +11	0.41	0.37	0.32	T + 23	-0.49	-0.49	-0.49
T + 12	0.35	0.3	0.25	T + 24	0 o,50	-0.5	-0.5

A perusal of Table 1 shows that by expressing a given temperature T in terms of mean temperature M and fraction of range R to be added or subtracted from the mean, one can get number of hours when temperatures will remain equal to or above the level of T from sunrise of day to sunrise of next day. The same are set out in Table 7.2 below.

Table 7.2 Number of hours equal to or greater than given value of T

T	Sunrise Hours			T	Sunrise Hours		
	5	6	7		5	6	7
Mean + 0.05R	11	12	11	Mean – 0.05R	13	12	13
Mean + 0.10R	11	11	11	Mean – 0.19R	14	14	14
Mean + 0.15R	10	10	10	Mean – 0.15R	14	15	14
Mean + 0,20R	10	9	9	Mean – 0,20R	16	15	15
Mean + 0.25R	8	8	9	Mean – 0.25R	16	16	17
Mean + 0.30R	7	8	7	Mean – 0.30R	17	18	17
Mean + 0.35R	7	6	6	Mean – 0.35R	19	18	18
Mean + 0.40R	5	5	5	Mean – 0.40R	19	19	19
Mean + 0.45R	4	4	3	Mean – 0.45R	20	21	21
Mean + 0.50 R	1	1	1	Mean – 0.50R	24	24	24

Monteith and Unsworth, 1990) and Rao et al., 1998) point out the need to take account of the time taken for Dew to dry off from crop foliage. The amount of Dew deposit per day even in winter in a per-humid region is low (Raman et al., 1973) and any relationship between LWD and any other derived parameter will automatically take care of the extra hours of LWD due to time taken by the Dew deposit to evaporate off. The above is borne out by the finding that DPD and NHRH > 90% models give estimates with good precision and accuracy of LWD relating to a turf grass and vineyard in a subtropical environment (Lulu et al., 2008).

7.9.2.1 Relative Humidity Factor

Daily Dew point temperature data is routinely available. Over a day, vapour Pressure, V.P. of air can be treated as a conservative parameter. The Saturation Vapour Pressure (SVP) at Dew point equals the vapour pressure of air. So for finding hours of temperature below dew point, the latter has simply to be expressed in terms of mean and range. Table 7.2 gives the number of hours when air temperature will be above Dew point in a 24 hour period and hence the hours when air temperatures will be below Dew Point i.e., duration of DPD.

If values above a specific R.H. are required, the concept of Equivalent Dew Point Temperature (EDP) can be used as follows. For example, for values above 75% find out the temperature at which V.P. of air will constitute 75% of SVP at that temperature, For example, if 13.4 °C is the Dew point V.P,. will be 15.36 and SVP required will be 15.36/0.80 = 19.2 mb which will be the SVP at 16.9 °C. Thus, 16.9 °C will be the EDP. Express EDP in terms of mean and range to obtain values of R.H > 75% as in case of routine Dew point temperature.

7.9.2.2 Crop Minimum Temperatures

The need for including the crop factor in LWD estimates (Sentelhas et al., 2006) arises from the fact that due to greater radiational cooling at night and transpirational cooling by day, the minimum and maximum temperatures will be lower than that recorded in a weather station. However, unlike the screen thermometer, crop foliage will absorb by direct solar radiation, which tends to raise the foliage temperature. The net result will be equality in maximum temperatures of crop and screen.

However, at night radiational cooling of crop foliage will be greater than for a screen thermometer A priori, the above will result in a lower minimum crop temperature. Cool and calm periods are conducive for dew deposition and in such weather, the depression of crop minimum below the screen minimum can vary from 1 to 5 °C across locations. The net result of a cooler crop minimum will be a longer duration of leaf wetness. For a mean temperature of 20 °C and Dew Point of 13 and ranges of 10, 16 and 20 °C the likely LWD for crop minimum depressions of 0, 1, 2, 3, 4, and 5 °C are indicated in Table 7.3.

Table 7.3 Increase in hours of LWD due to cooling of leaf foliage below screen minimum

Temperatures °C					Depression of Crop Minimum °C					
Max.	Min.	Mean	Range	D.P.	0	1	2	3	4	5
25	15	20	15	13	0	0	0	4	6	7
28	12	20	16	13	3	5	6.5	7	8	9
30	10	20	20	13	6	7	7.5	8	9	9

The above postulations show how routinely available temperature data can be used to provide the inputs required in disease forewarning models and the necessity for taking account of the climatologically appropriate values of depressions of grass minimum temperature at a given location and period in computing LWD.

7.10 Mid Seasonal Advisories

Advice for initiation of control operations against a pest or disease is crucial as pest is most vulnerable at the egg laying and egg hatching stages and a disease organism is best controlled at the sporulation stage. Pests and diseases after initiation have peaks of infection and initial applications of insecticides and fungicides may not be able to totally prevent setbacks due to them. Thus, mid-seasonal advisories for resumption of control action may be required. There are intra-seasonal variations in severity of pests and diseases, For example, Sesbania thrips on groundnut showed one peak in the rainy season, two peaks in the winter season and three peaks in the summer at Junagadh, Gujarat India (Prasad et al., 2008). At Pantnagar, India Singh and Sachan (1993) found four peaks of catches of male moths of Spodoptera litura in groundnut in pheromone traps. Gedia et al., (2008) found three peaks in the oviposition and egg masses of spodoptera litura traps in traps in cotton fields of Gujarat. Thus, issue of mid-seasonal advisories will be comparatively more difficult than issue of advice for commencement of spraying.

In the case of some diseases, the above problem is sought to be overcome through the concept of Disease Severity Value, DSV (Madden et al., 1978). In this, after a specified initial period, DSVs are commenced to be accumulated and each day a DSV is calculated as per a formula such as

$$DSV = DAP/ (45(Y+X)$$

Where DAP = Days after planting, Y = Hours of Relative Humidity of 95-100% at temperature of 20 $^{\circ}$C divided by 8 and X = Hours of leaf wetness.

The ratings are accumulated to a threshold value that calls for initiation of action. After the action, DSV is set to zero and the accumulations commenced till they reach an assigned value calling for another spray and the process is repeated till end of the crop period. For example, in the TOMCAST disease management of

Tomatoes, the first fungicide spray is applied when DSV reaches a value of 35 and additional sprays are applied when 29 DSVs are accumulated (Gillespie et al., 1993).

If adequate data are available, similar concept of Pest Severity Value can be developed. For example Singh et al., (2009) found that the maximal possible population of larvae of the Lucerne Weevil, Hypera postica (Gyllenhal) was 500 per square foot and that the weevil population Y could be predicted from the equation:

$$Y = 110.46 + 0.93049 \, X - 0.000 X^2$$

Where X is the accumulated degree days above a base of 9 °C from 1st December. The above curvilinear relationship could be used to assign Biofix values for trap catches in terms of population level values for initial and subsequent sprayings.

In case of fruit trees which are attacked by more than one generation of a pest, Boifix criteria of trap catches coupled with degree day accumulations are used to time both initial and subsequent sprays as needed. For example, for the Codling Moth, Cydia pomennella the first Biofix is the 5th moth in the trap. The first spraying is advised after accumulation of 259 degree days above a base of 50 °F. After 1000 day degree accumulations from first Biofix, a watch is kept for an increase in number of trap catches. Spraying is resorted to after accumulation of 250 day degrees from the 2nd Biofix. After 1000 day degree accumulations of 2nd Biofix a watch is kept for increase in number of trap catches and spraying resorted to after 250 day degree accumulations from the 3rd Biofix.

A similar procedure is adopted in the case of the Oriental Fruit Moth. Here, the first Biofix is the occurrence of more than one larva per trap. First spraying is done after accumulation of 200 day degrees above a base of 42 °F. The second Biofix is the date of emergence of adults. Second spraying is done after accumulation of 200 day degrees from the 2nd Biofix. The 3rd Biofix is given by the trapping of more than 8 moths per trap per week and 3rd spraying is done after accumulation of 200 day degrees above a base of 42 °F.

7.11 Effectiveness of Forewarnings

Forewarnings based on current crop position and expected weather are most effective when there is a time lag between the onset of favourable conditions and the manifestation of pest or disease affliction. Examples of such types are given by (i) Starr (1943; 1950) regarding black rust of wheat in Oklahoma, USA (ii) Miller (1937) and Stevens and Ayres (1940) regarding downy mildew of Tobacco in Southern USA. Wellington (1954) has drawn attention to the weather in the preceding year determining whether the Tent Caterpillar or the Spruce Budworm would become important in the succeeding year in forest areas. The other situation is when an initial inoculum is detected in insect or spore traps and the organism's phenological development is amenable to calculation by the accumulated degree days above a base value. An eminent example is that of the Codling Moth, Cydia pomenella (Vasev, 1963). Another situation is when conditions are ripe and await only the weather trigger. Example is Soybean in the reproductive phase with viable spores of rust and expected cool wet weather (Quales and Young, 2006), Rice in the tillering or heading stage with viable spores waiting for minimum temperatures to go below 23°C and the R.H. to reach 90% (Padmanabhan et al., 1971) or for the Dynamic Cumulative Weather Based Disease Severity Index to reach 0.50 (Yella Reddy et al., 2006) for blast to occur are other examples. Use of regressions containing both weather and crop factors with a lead time of one week offer scope for effectiveness. Last but not the least, effectiveness of forewarnings is maximal over large mono-crop areas and least in situations of variegated cropping over fragmented and small holdings.

7.12 Weather and Chemo-Control

Weather conditions have a great bearing on the mode and effectiveness of application of biocides against pests and diseases. Weather at the time of operation determines if the chemical is to be sprayed or dusted, Weather immediately after application of a biocide determines its effectiveness. For example, following an application, a rain-free period of 18 hours and 6 hours respectively are deemed desirable for non-systemic and systemic biocides respectively. Strong wind tends to blow-off dusted chemicals. The winds must be light and smooth and the direction of application of the chemical must be perpendicular to the wind direction.

Under sodden ground conditions, over forests, plantation crops and of crops in undulating terrain, for control of insects that move with the wind and as social benefit measure, aerial applications of biocides have to be resorted to. The meteorological influences on and the weather requirements for aerial and ground-based applications are quite different (Bourke et al., 1960). In aerial applications the concentration of the chemical is higher and of a finer size and the release is from a greater height and absence of thermal winds and low evaporative power of air are a must. In ground-based application to detect low level turbulence, setting off smoke bombs or small trash fires for determining the time taken for diffusion of the smoke into the air is advocated (Bourke et al., 1960). Visibility at the crop canopy level becomes important for pilotage of the aircraft. The weather requirements for ground-based and aerial applications are set out the WMO Guide for Agricultural Meteorological Practices.

7.13 Weather in Biological Control

The concept of Biological control of pests and diseases has been mentioned in the section on natural enemies. In this a natural enemy identified through careful surveys is imported into a region to control the virulence of the concerned pest/disease. In such a case it stands to reason that an evaluation of the climatology of the local and alien habitats would help time the introductions to ensure minimal mortality of the natural enemy during transport and maximal increase after release. Delineated Homoclimates as set out in an earlier chapter can help decide on the usefulness or otherwise of exotic introduction of the envisaged natural enemy.

7.14 Real-Time Forewarning of Pests and Diseases

Several simulation models for forewarning, mostly of diseases, are reported to be in operation in developed countries. In India, forecasting models relate to leaf blast of Rice (Manibushan Rao and Krishnan, 1991), Alternaria blight and Aphid (Lipaphis erysimi) in Brassica (Chattopadhyay, 2005 a and b), Powdery Mildew on Brassica (Desai et. al., 2004) Pod Borer (Helicoverpa armigera) on Pigeon Pea (Dhar et. al., 2007). Aphid, (Myzus Persicae (Sulzer)) on Potato (Trivedi et. al., 1999). The Indian Agricultural Statistical research Institute, IASRI in collaboration with the Centre for Dryland Agriculture in India, CRIDA and International Crop Research Institute for Semiarid Tropics, ICRISAT had developed methodologies (Mehta et al., 2001; Ranjana and Mehta, 2007) for forewarning of Alternaria Blight, White Rust, Powdery Mildew, Mustard Aphid, various Boll worms and white Fly of Cotton, late leaf blast and rust of groundnut, Shoot and Top Borers of Sugarcane, Pyrilla and Gall Midge of Rice etc. The above have been reported successfully used by several institutions. Accuracy of medium range weather forecasts have improved and the same are being increasingly used in the framing and issue of crop cum weather based agronomic advisories. The need for use of the concepts of ETL and trap catches for insects and spores as part of the IPM and IDM programmes are being increasingly welcomed. Validation of routinely available data to derive inputs needed in disease and pest forewarning models and organisation of a network of aero-entomological and aero-mycological stations for trapping insects and spores and study the data thereof as forewarning tools and to elucidate weather relations of incidence of pests are called for. With the requisite infrastructure in place the available methodologies offer a very good scope for forewarning of pests and diseases on a real-time and operational basis.

References

Adalberto, T.M. 1981. An exponential model of the curve of mean monthly hourly temperatures. Atmosfera. 4: 139-144.

Arora, R.K.; Sharma, K.K. and Bombawale, O.M. 1999. Management of potato Late Blight by changing planting date in North Western Plains. Jl. Mycol. Plant Pathol. 29: 355-358.

Aylor, D.E. 1986. A framework for examining inter-regional aerial transport of fungal spores Agric. and Forest Meteorol. 38: 263-288.

Bal, S.K. et al. 2008. Agroclimatic regression model for forecasting Ascochyta blight of chick pea in Punjab. Jl. of Agrometeorol. 1: 201-204.

Bourke, P.M.A. 1955. The forecasting from weather data of potato blight and other plant diseases and pests. Tech, Note No. 10. World Met. Org. (WMO), Geneva, 48 pp.

Bourke, P.M.A. et al. 1960. Meteorological Service for aircraft employed in agriculture and forestry. Tech, Note No, 32, WMO, Geneva, 32 pp.

Broadbent, L. 1950. The micro-climate of the potato crop. Jl. Royal Met. Soc. 76: 439-454.

Chakravarty, N.V.K. and Gautam, R.D. 2004. Degree-day based forewarning system for mustard aphid. Jl. of Agrometeorol. 6: 215-222.

Chattopadhyay, C. et. al. 2005a. Epidemiology and forecasting of Alternaria blight of oilseed Brassica in India – A case study. Zeitsschrift fur Pflanzenkrankheiten und Pflanzenschutz. (Jl. of Plant Diseases and Protection). 112: 351-365.

Chattopadhyay, C. 2005b. Forecasting of Lipaphis erysimi on oilseed Brassicas in India – A case study. Crop Protection, 24: 1042-1053.

Chaudhari, G.B. et al. 1999. Effect of weather on activity of cotton boll worms in Gujarat, Jl. of Agrometeorol. 1: 137-142.

Chaudhary, J.P. and Sharma, A.K. 1982. Feeding behaviour and larval population levels of Helicoverpa armigera causing economic damage to gram crop. Haryana Agric. Univ. Jl. of Res. 12: 462-466.

Dang. K. and Doharey, K.I. 1971. Effect of temperature and humidity on development of the sorghum shoot borer, Chilo zonellus. Investigations on insect pests of sorghum and millets, 1965-70. Final Technical Report. of Ind. Agricl. Res. Inst. 138-144 pp.

Darpoux, H. 1943. Les bases scientifiques des avertissements agricoles. Annales Des Epiphyties et de Phytogenetique, 9: 177-208.

Das, D.K.; Trivedi, T.P. and Srivastava, C .P. (2001). Simple rule to predict attack of Helicoverpa armigera on crops growing in Andhra Pradesh. Ind. Jl. Agric. Sci. 71: 421-423.

Das, D.K. 2013. Personal Communication.

Delponte, E.M.; Godoy, C.V.; Li. X. and Yang, X.B. 2006. Predicting severity of Asian soybean rust with empirical rainfall models. Phytopathol. 96: 797-803.

Desai et. al., 2004. Brassica juncea powdery mildew epidemiology and weather-based forecasting models for India – A case study, Zeitschrift fur Pflanzenkrankheiten und Pflanzenschutz. (Jl. of Plant Diseases and Protection). 111: 429-438.

De Villiers, G.D.B. 1966. Plant disease, insects and weather. Proc. WMO Agromet Training Seminar, Melbourne, Australia. 667-693 pp.

Dhaliwal, L.K. and Hundal, S.S. 2004. Use of weather variables in prediction of Mustard Aphid, Lipaphis Ersymi incidence. Jl, of Agrometeorol. 6, Spl. Issue: 136-140.

Dhaliwal, L. K; Koona, B.S,; Singh, J. and Sohi, A.S. 2004. Incidence of Helicoverpa armigera (Hubner) in relation to meteorological parameters under Punjab Conditions. Jl. of Agrometeorol. 6, Spl. Issue : 115-119.

Dhar, V. et. al., 2007. Prediction of pod borer, (Helicoverpa armigera) infestation in short duration pigeon pea (Cajanus cajan) in Central Uttar Pradesh. Ind. Jl. Agric. Sci. 77: 701 – 704.

Dhawan, A.K.; Anega, A.; Singh, J. and Sarika, S. 2009. Population dynamics of different pests on Bt cotton vis-a-vis meteorological parameters in Punjab. Jl. of Agrometeorol,. 11: 180-182.

Dhawan, A.K. et al; 2007. Incidence and damage potential of mealy bug (Phenacocus solenopsis, Tinsley) on cotton in Punjab. Ind. Jl. Ecol. 34: 166-172.

Douglas, D.R. and Grasshopp, M.D. 1974. Control of early blight in Eastern and South Central Idaho. Amer. Jl. Potato Res. 51: 361-368.

Dubey, R.C. and Venkataraman, S. 1980. A note on the likely areas of establishment of Japanese Beetle (Popillia japanica, Newman) in India. Mausam, 31, 175-177.

Dubey, R.C. and Yadav, T.S. 1980. Sorghum shoot fly (Atherigona Soccata, Rondani) incidence in relation to temperature and humidity. Ind. Jl. Entomol. 42: 273-274.

Emerson, M.D.P. and Esker, P.D. 2008. Meteorological factors and Asian Soybean Rust epidemics- a system approach and implication for risk assessment. Sci. Agric. 65: 88-07.

Farrington, J. 1977. Economic threshold of insect population in present agriculture. A question of applicability. Pest Articles and News Summaries, 23: 143-148.

Franc, G.D.; Harrison, M.D. and Lahman, L.H. 1988. A simple day-degree (DD) model for initiating chemical control of potato early blight in Colorado. Plant Dis. 72: 851-854.

Francl, L.J, and Panigrahi, S. 1997. Artificial neural network models of wheat leaf wetness. Agric. and Forest Meteorol. 88: 57-65.

Fry, W.E. and Apple, A.E. 1986. Disease management implications of age-related changes in susceptibility of potato foliage to Phytopthora infestans. Am. Potato Jl. 63: 47-56.

Gedia, M.V.; Vyas, H.J.; and Acharya, M.F. 2008. Weather-based monitoring of male moths in pheromone traps and oviposition of Spodoptera litura on cotton in Gujarat. Jl. of Agrometeorol. 10: 81-85.

Gillespie, T.J.; Srivastava, B. and Pitbadlo, R.K. 1993. Using operational weather data to schedule fungicide sprays on tomatoes in Southern Ontario. Canadian Jl. Applied Meteorol. 22: 567-573.

Gleason, M.L. et. al., 1994. Development and validation of an empirical model to estimate the duration of the dew periods. Plant Disease, 79: 1011-1016.

Grace, A.D. et al. 2009. Integrated management of Helicoverpa pod borer on Pigeonpea at Nalgonda and Mehboobnagar districts of Andhra Pradesh. Ind. Jl. Environ. and Ecoplan. 16: 1-4.

Gregory, P.H. 1973. The Microbiology of the Atmosphere. Wiley, New York. 377 pp.

Gupta, B.D. and Kuslhreshta, J.P. 1957. Control of Johnson grass (Sorghum halpense) – An alternate host of sugarcane stalk borer Chilotraea auricilia. News Letter Series II. Entomological research. Ind. Instt. Sugarcane Res. Lucknow, India, 15-16.

Gupta, M.P.; Nayak, M.K. and Srivastava, A.K. 2009. Studies in seasonal activity of white fly Bemisia tabaci, genn) population and its association with weather parameters in Bundelkhand zone of Madhya Pradesh. Jl. of Agrometeorol. 11: 175-179.

Gupta, R.B.V.L. et al. 2003. Understanding *Helicoverpa armigera* Pest Population Dynamics related to Chickpea Crop Using Neural Networks. Third IEEE International Conference on Data Mining, Melbourne, Florida, 19-22.

Harrison, M.D,; Livingston, C.H. and Oskima, N. 1965. Control of potato early blight in Colarado. H. Spore traps as a guide for initiating application of fungicides. Amer, Potato Jl. 42: 333- 341.

Hoffmann, M.P.; Wilson, L.T.; Zalom, F.G. and Hilton, R.J. 1991. Dynamic sequential sampling plan for Helicoverpa zea (Lepidoptera Noctuidae) eggs in processing tomatoes-Parasitism and temporal patterns. Environmental Entomol. 20: 1005-1012.

Howell, J. T. and Niven, L.G. 2000. Physiological development time and zero development of the codling moth (Lepidoptera: Torticidae). Environmental Entomology, 28: 766-772.

Jeyrani, S.; Sridharan, S. and Sadahatulla, S. 2000. Inter-relationship between light trap catches and field population of Brown Plant Hopper, Nilparvata lugens (stal). Jl. Insect Environ. 6: 59-60.

Jha, R. C. and Bisen, R.S. 1994. Effect of climatic factors on the seasonal incidence of the pink boll worm on cotton crop. Annals of Plant Protection Science. 2: 12-14.

Jhagirdar, S; Venkatesh, H and Jamadar, M.M. 2001. Influence of antecedent weather variables on powdery mildew of grapes in Northern Karnataka. Abstracts, National Seminar on Agrometeorological Research for Sustainable Agricultutral Production, Anand, Gujarat India 112 pp.

Jorhar, et al, 1992. A biometeorological model for forecasting Karnal bunt disease of wheat Plant Dis, Res. 7: 204-209.

Joshi, L.M. 1963. Development of rusts: Effects of temperature on infection of wheat varieties with Puccinia graminis tritici. Ind. Phytopathol. 15: 290-291.

Joshi, L.M. and Palmer, L.T. 1973. Epidemiology of stem, leaf and stripe rusts of wheat in Northern India. Plant Dis. Reptr. 57: 8- 12.

Jotwani, M.G.; Marwaha, K.K.; Srivastava, K.M. and Young, W.R. 1970. Seasonal incidence of shoot-fly (Atherigons soccata) in jowar hybrids at Delhi. Indian Jl. Entomol. 32: 7-15.

Kalra, A.N. and Sharma, N.C. 1863. Occurrence of the shoot borer. Chilotraea infuscattellus (Snell) as cane borer in the Sriganganagar area of Rajasthan. Ind, Jl. Sug. Res & Develop. 7: 192-194.

Kaur, G.; Kaur, S. and Hundal. S.S. 2006. Dispersal of secondary inoculum of Tiletia indica causing karnal Bunt of wheat in relation to weather parameters, Jl.of Agrometeorol. 8: 215-222.

Khan, S.A.; Choudhari, S. and Jha, S, 2008, Weather based forecasting of mustard aphid (Lipaphis erysimi, (Kalt), Jl. of Agrometeorol. Spl. Issue Part 1, 186-189.

Kharbede, S.B.; Wayal, C.B. and Navale, P.A. 2007. Integrated pest management in cotton. Ind, Jl,   Environ. and Ecoplan, 14: 567-570.

Kim, K.S.; Taylor, S.E. and Gleason, M.I. 2004. Development and validation of a leaf wetness model using a fuzzy logic system. Agric. and Forest Meteorol. 127: 53-64.

Kler, D.S.; Kaur, N. and Uppal, R.S. 2005. Soil and groundwater pollution by agrochemicals- A review. Ind, Jl. Environ and Ecoplan, 7: 285-294.

Krishnaiah, N.V. et al. 2006. Population dynamics of rice brown plant hopper (Nilparvate lugens) in Godavari delta in Andhra Pradesh state. Ind. Jl. Plant Protection, 34: 158-164.

Krupa, S. et al. 2006. Introduction of Asian soybean rust urediospores into the Midwestern United States- A case study. Plant Disease 90: 1254-1259.

Kumar, G. and Chakrvarty, N.V.K. 2008. A simple weather based forewarning model for white rust in Brassica. Jl. of Agrometeorol. 10: 75-80.

Kumar, V.; Kaushik, C.D. and Gupta P.P. 1995. Role of various factors in development of white rust disease of rapeseed mustard. Ind. Jl. Mycol. Plant Pathol. 25: 145-148.

Kundu, G.G.; Kishore, P. and Jotwani, M.G. 1971. Seasonal incidence of shoot-fly, Atherigona soccata Rondani at Udaipur (Rajasthan). Investigations on insect pests of sorghum and millets, 1865-1970. Final technical report of Indian Agricl. Res. Instt. Delhi. 131-137 pp.

Lulu, J. et al. 2008. Estimating leaf-wetness duration over turf grass and in a Nigara Rosada vine yard in a subtropical environment. Scientia Agricola (Piracicaba Braz,). 65: 10-17.

Madden, L., Pennypacker, S. P., and McNab, A. A. 1978. FAST, a forecast system for *Alternaria solani* on tomato. Phytopathology. 68: 1354-1358.

Manibhushanrao, K. and Krishnan P. 1991. Epidemiology of blast (EPIBLA): a simulation model and forecasting system for tropical rice in India. Rice blast modeling and forecasting. IRRI Manila Philippines pp. 31-38.

Mavi, H.S. et al. 1992. Forecasting Karnal bunt of wheat - a meteorological method. Cereal Res. Communications. 20: 67-74.

Mehta, K.C. 1940. Further studies on the cereal rusts in India. Monograph No.14, Ind. Counc. Agric. Res, ICAR New Delhi, India. 224 pp.

Mehta, K.C. 1952. Further studies on the cereal rusts in India. Monograph, No. 18, ICAR, New Delhi India. 165 pp.

Mehta, S.C.; Ranjana, A. and Kumar, A. 2001.

ww.iasri.res.in/Sovenior/Article_07.pdf pp 67-77.

Metange, K.K. et al. 2004. Effect of soil temperature on adult population of Helicoverpa armigera, Hubner. Jl. of Agrometeorol. 6: 278-279.

Miller, P.R. 1959. Plant Disease Forecasting in Plant Physiology-Problems and Progress Ed. C.S. Holton et al. Wisconsin Press Madison, U.S.A. 557-565 pp.

Miller, P.R. 1937. January temperatures in relation to the destruction and severity of downy mildew of tobacco. Plant Dis, Reptr. 21: 260-266.

Mishra, M.D.; Raychaudhuri, S.P.; Everett, T.R. and Basu, A.N. 1973. Possibilities of forecasting outbreaks of Tungro and yellow dwarf of rice in India and their control. Proc. Symp. Epidemiology, forecasting and control of plant diseases. Ind. Nat. Sci. Acad, Bull. No. 46: 352-356.

Monteith, J.L. and Unsworth, M.H. 1990. Principles of Environmental Physics. Second Edition. Elsevier Publishing co. New York, U.S.A. 291pp.

Nagarajan, S and Singh, H, 1973. Satellite television cloud photography as a possible tool to forecast plant disease spread. Curr. Sci. 42: 273-274.

Nandagopal, V.; Prasad, T. V. and Gedia, M.V. 2006. Population dynamics of leaf-eating caterpillar, Spodopters litura (Fab) in relation to weather parameters in groundnut in South Saurashtra region. Jl. of Agrometeorol. 8: 60-64.

Nandagopal, V.; Prasad, T.V.; Gedia, M.V. and Makwana, A.D. 2008. Influence of weather parameters on the population dynamics of Sesbnia thrips, Jl. of Agrometeorol. 10: 173-177.

Nandal, D.P.; Om, H. and Dhiman, S.D. 1969. Efficiency of herbicides applied alone and in combinatiom against weeds in transplanted rice. Ind, Jl. Weed Sci. 31: 239-242.

Nandihalli, B.S.; Patil, B.V.; Somasekhar, S. and Hugan, P. 1989. Influence of weather parameters on the population dynamics of Spodoptera litura (Fb) in pheromone and light traps. Jl. Agric. Sci. 2 : 62-67.

Nayban, G. and Chowdhury, P. 2002. Cited by Khan et al; 2008.

Nel, J. C . Cited by Devilliers, 1966.

Nield, R. 1967. Maximum–minimum temperatures as bases for evaluating thermoperiodic response. Mon. Wea. Rev. 95: 583-584.

Ogwaro, K. and Kokwaro, E.D. 1981. Development and morphology of the mature stages of the sorghum shoot fly. Atherigona Soccata, Rondani. Insect Sci. Application, 1: 365-371.

Padmaja, P.G.; Prabhakar,; M.; Yella reddy, D. and Prasad, Y.G. 2005. Influence of weather on Sorghum shoot fly, Atherigona Soccata (Rondani) and models for forewarning their incidence, Jl. of. Agrometeorol. 7: 51-58.

Padmanabhan, S.Y. 1965. Studies on forecasting outbreaks of diseases of rice: Influence of meteorological factors on blast incidence at Cuttack. Proc, Ind. Acad. Sci. 62: 117-120.

Padmanabhan, S.Y and Ganguly, D. 1954. Relation between the age of the rice plant and its susceptibility to Helminthosporium and blast diseases. Proc. Plant Sciences, 39: 44-50.

Padmanabhan, S.Y.; Chakrabarti, N.K. and Rao, K.V.S.R.K, 1971. Forecatsing and control of rice diseases. Proc. Nat. Sci. Acad. 37. Part B. 423-429.

Patel, H.R. et al. 2004. Early blight management and its effect on tuber yield of two potato cultivars under varied environmental conditions, Jl. of Agrometeorol, 6: 229-232.

Patel, R.S. and Yadav, D.N. 2004. Conservation of orthopod natural enemies by interspersing of maize   in   cotton   pests. International Symp. on Strategies for sustainable cotton production - A global vision. UAS, Dharwad, India. 23-25 Nov. pp 270-273.

Parton, W.J. and Logan. J.A. 1981. A model for diurnal variation in air and soil temperature, Agric. Meteorol. 23: 205-216.

Pedigo, L.P. 1991. Entomology and Pest management. Macmillan Publishing Co. New York. Pages 107-119.

Pierce, W.D. 1934. At what point does insect attack becomes damage. Entomological News. 45: 1-4.

Pierce, W.D.; Cushman, R.A, and Hood, C.E. 1912. The insect enemies of the cotton boll weevil. U.S. Dept. Agri. Bur. Entomol. Bull. 100: 1-99.

Pivona, S. and Yang, X.B. 2004. Assessment of potential year round establishment of soybean rust throughout the world. Plant Disease. 88: 523-529.

Ponnaiya, B.W.X. 1951. Studies on the genus sorghum. Field observations on sorghum resistance to the   insect   pest   - Atherigona indica. Madras Univ, Jl. 21: 105.

Prabhakar, M.; Singh.Y.; Singh, V.S. and Singh, V.P. 1998. Economic injury levels of Helicoverpa armigera in Chickpea as influenced by irrigation. Ind. Jl. Entomol. 60: 109-115.

Pradhan, S, 1946. Idea of a Biograph and Biometer. Proc. Nat. Instt. Sci. India. 12: 301-314.

Prasad, L.T.V.; Nandagopal, V. and Gedia, M.V. 2008. Seasonal abundance of Sesbania thrips, Caliothrips indicus, Bagnall in groundnut,. Jl. of Agrometeorol. Spl. Issue Pt.1. 211-214.

Quales, R.V. and Young, X.B. 2006. Spore trap help researchers watch for Soybean rust. Integrated Crop management. IC – 496(16): 185 pp.

Qureshi, Z.A.; Ahmad, N. and Hussain, T. 1993. Pheromone trap catches as a means of prediction of damage by Pink Boll worm larvae in Cotton. Crop Protection. 12: 597-600.

Raman, C.R.V.; Venkataraman, S. and Krishnamurthy, V. 1973. Dew Over India and its Contribution to Winter- crop Water Balance. Agric. Meteorol. 11: 17-35.

Ranga Rao, G.V.; Wightman, J.A. and Ranga Rao, D.V. 1991. The development of a standard pheromone trapping procedure for Spodoptera litura (F.) (Lepidoptera: Noctuidae) population   in groundnut (Arachis hypogaea L.) crops. Tropical Pest Management. 37: 37-40.

Ranga Rao, G.V. et. al., 2008. Influence of weather factors on insect pests and diseases on groundnut in India: A Case study, Jl. of Agrometeorol. Spl. Issue Part 1. 502-507.

Ranjana, A. and Mehta, S.C. Weather-based forecasting of yields, pests and diseases -IASRI Models. Jl. Ind. Soc, Agric. Statistics 61: 255-263.

Rao, P.S.; Gillespie, J.J. and Schaafsma, A.W. 1998. Estimating wetness duration on maize ears from meteorological data. Canadian Jl. Soil Sci. 118: 149-154.

Rao, Y.P. and Srivastava, D.N. 1973. Application of phage in investigations on epidemiology of bacterial blight disease of rice. Proc. Symp. on Epidemiology, Forecasting and Control of Plant Diseases. Indian National Sci. Academy Bulletin. 46: 313-321.

Raodeo, A.K. and Muqum, A. 1971. Effect of climatic factors on the population of jowar stem maggot, Atherigona soccata Rondani, at Parbhani. Coll. Agric. Mag. Parbhani. 12: 12-16 pp.

Rathi. A.S.; Gupta, P.P.; Grewal, R.P.S. and Ram Niwas, 1999. Environmental Factors in relation to development of foliar diseases of sorghum. Forage Res. 25: 175-178.

Riedl. H.; Croft, B.A. and Howitt, A.J. 1976. Forecasting codling moth phenology based on pheromone trap catches and pysiological-time models. Can. Entomol. 108: 449-460.

Robertson, G.W. 1983. Ed. Guidelines on crop weather models Pub. World. Met. Organisation. Personal Communication. 202 pp.

Roy, S.K. and Baral, K. 2004. Influence of weather parameters on arrival and dispersal of Mustard Aphid. Lipaphis erysimi (Kaltenbach). The Ekologia. 2: 123-127.

Roy, M., J. Brodeur, and C. Cloutier. 2002. Relationship between temperature and developmental rate of *Stetho-rus punctillium* (Coleoptera: Coccinellidae) and its prey *Tetranychus mcdanieli* (Acarina: Tetranychidae). Environ. Entomol. 31: 177-187.

Sadasivam, T.S.; Suryanarayanan, S. and Ramakrishnan, L. 1963. Influence of temperature on rice blast disease. In; The rice disease. John Hopkins Press, London, pp 163-171.

Sadler, E.J. and Schroll, R.E. 1997. An empirical model of diurnal temperature patterns, Agron. Jl. 89: 542-548.

Saha, G. et. al., 2008. Studies on date of initiation of late blight of potato based on disease progress curve, Jl. of Agrometeorol. 10: 170-174.

Saharan, G.S.; Kaushik, C.D.; Gupta, P.P. and Tripathi, N.N. 1984. Assessment of losses and control of white rust of mustard. Ind. Phytopathol. 37: 397.

Salmon, S.C. 1951. Forecasting the occurrence of diseases and insects in Japan. Plant Dis. Reptr. 35: 251-254.

Samui, R.P.; Chattopadhyay, N,; Sabal, J.P. and Balachandran, V.P. 2004. Weather based forewarning models for major pests of rice in Pattambi Region (Kerala) Jl. of Agrometeorol. Spl. Issue. 6. 105-114.

Samui R.P.; Chattopadhyay, N,; Sabal, J.P. and Balachandran, V.P. 2008. Population dynamics of   stem borer in relation to inter and intra seasonal variation of weather and operational rice protection at Pattambi, Kerala, Jl. of Agrometeorol. 10. Spl, issue Part 2. 512-519.

Saravanan, T.; Ragavan, T.; Venkataraman, N.S. and Subramanian, V. 2008. Degree-day based method for predicting the occurrence of Ersyphi polygoni in Vigna mungo, L. Jl. Of Agrometeorol. Spl Issue Pt. 1, 205-210.

Sardar Singh, 1968. Plant protection problems of cropping patterns. Proc,. Symp. Cropping Patterns in India. Ind, Council Agric. Res., ICAR, 153-161.

Schenck, N.C. 1968. Fungicidal control of watermelon downy mildew and its relationship to first infection in the field. Plant Dis. Rep. 52: 979-998.

Sentelhas, P.C. et. al., 2004. Operational exposure of leaf wetness sensors. Agric. & Forest Meteorol. 141: 105-117.

Sentelhas, P.C. et. al., 2006. Evaluation of a Penman-Monteith approach to provide "reference" crop canopy leaf wetness duration. Agric. & Forest Meteorol. 141: 105-117.

Sentelhas, P.C. et. al., 2008. Suitability of relative humidity as an estimator of leaf wetness duration. Agric. and Forest Meteorol. 148: 392-400.

Shamim, M. et. al., 2009. Effect of weather parameters on population dynamics of green leaf hopper and white backed plant hopper in paddy grown in middle Gujarat. Jl. of Agrometeorol. 11: 172-174.

Sharma, P.K. and Kashyap, N.P. 1998. Estimation of losses in three different cruciferous oil seed Brassica crops due tomato aphid complex in Himachal Pradesh (India), Jl. Entomol. Res. 22: 337-342.

Shoemaker, P.B. and Lorbeer, J.W. 1969. Timing protection spray initiation to control onion leaf blight. Phytopathology, 67: 402-409.

Shoemaker, P.B. and Lorbeer, J.W. 1977. Timing of fungicide application to control Botrytis blight epidemics on onions. Phytopathology, 67: 409-414.

Singh, S.P. and Jalali, S. K. 1997. Management of Spodoptera litura (Fabricius) (Lepidoptera: Noctuidae) Proc, National Scientists Forum on Spodoptera litura. ICRISAT, Hyderabad, India. 27-65 pp.

Singh, H.; Rohilla, H.R.; Kalra, V.K. and Yadav, T.P. 1984. Responses of Brassica varieties sown on different dates to the attack of Mustard Aphid, Lipaphis ersymi (Kalt). Jl. Oilseeds Research. 1: 49-56.

Singh, J. P. and Lather, B.P.S. 1989. Monitoring of pink boll worm moths and larvae. Ind, Jl.of Plant Protection. 17: 199-204.

Singh, J.B.; Pandey K.C.; Saxena, P. and Behari, P. 2009. Degee-day model for development and Incidence of lucerne weevil Hypera postica (Gyllenhal) in Central India. Curr. Sciemce 96: 1578-1580.

Singh. J.B.; Saxena. P. Pandey. K.C. and Behari, P. 2008. Influence of weather parameters on zonate leaf spot) (Glocercospora sorghi) development on Sorghum, Jl. of Agrometeorol. Spl. Issue Pt.1: 186-189.

Singh, K.N. and Sachan, G.E. 1993. Assessment of the use of pheromone traps in the management of Spodoptera litura F. Indian Jl. Plant Protection, 21: 8-13.

Srivastava, N.N. et. al., 2010. Studies on diurnal air temperature pattern from daily maximum and minimum by estimating the parameters of sinusoidal and exponential models on weekly basis under semi arid climate of Hyderabad. Jl. of Agrometeorol, 12: 8-14.

Starr, C.K. 1943. The divisive influence of late winter weather in wheat leaf rust epiphytotics. Plant Dis. Reptr. Suppl. 143. 133-144.

Starr, C.K. 1950. Validity and values of plant disease forecasting. Plant Dis. Reptr. Suppl. 190: 5-8.

Stern,V, et. al., 1959. The integrated control concept. Hilgardia, 29: 81-101.

Stevens, N.E. and Ayres, J.C. 1940. The history of tobacco downy mildew in United States in relation to weather conditions. Phytopathol. 30: 684-688.

Subbiah, K.S. and Mohamed Ibrahim, P.A. 1968. A preliminary study on the seasonal occurrence and abundance of Shoot-fly, Atherigona indica Mallock with reference to highly susceptible variety of sorghum at Kovilpatti. Madras Agric. Jl. 55: 88-90.

Sundaramurthy, V.T. and Chitra, K. 1996. Integrated pest management in cotton. Ind. J. Plant Protection, 20: 1-17.

Sulaiman, M. and Agashe, N.G. 1965. Influence of climate on Tikka Disease of groundnut in Maharashtra. Ind. Oilseeds Jl. 9: 176-179.

Sutton, T.B. and Jones, A.L. 1976. Evaluation of four spore traps for monitoring discharge of ascospores of Venturia inequalis. Phytopathol. 66: 453-456.

Tamaki, Y.; Noguchi, H. and Yushima, T. 1973. Sex pheromone of Spodoptera litura (F)) (Lepidoptera : Nooctuidae). Isolation, identification and synthesis. Appl. Entomol. Zool. 8: 200-203.

Thorold, C.A. 1952. The ephiphytis of Theobroma Cacos in Nigeria in relation to incidence of Black Pod disease (Phytopthora palmivora). Jl. Ecol. 40: 125-142.

Trivedi. T.P.; Jain, R.C.; Mehta, S.C. and Dhar, L.M. 1999. Developments of forewarning systems for aphid, Myzus Periscae (Sulzer) on Potato. Annual Report NCIPM, New Delhi, India. 43-49.

Trivedi, T.P. et. al., 2005. Monitoring and Forecasting of *Heliothis/Helicoverpa* Populations. In Heliothis/Helicoverpa Management: Emerging Trends and Priorities for Future Research (Ed. Sharma, H.C.). Oxford and IBH Publishing Co. New Delhi. 119- 140 pp.

Usman, S. 1968. Preliminary studies on the incidence of shoot-fly on hybrid jowar under differential sowing. Mysore Jl. Agric. Sci. 2: 44-58.

Uvarov. B.P. 1961. Insects and Climate. Trans, Entomol . Soc. London. 79: 1-24.

Vadodaria, M.A. et. al., 1997. Integrated management of cotton boll worms in Gujarat. Ecological Agriculture and sustainable Development. Vol. II. 282-285 pp.

Varma, N.R.G.; Bhanu, K.V. and Raji Reddy, D. 2008. Forecasting population of brown plant hopper (Nilparvata lugens (Stal). Jl. of Agrometeorol. Spl. Issue Pt. 1, 197-200.

Vasev, A. 1963. Cited by De Villiers, 1966.

Venkataraman, S. 2002, Tabular aids for computation of derived agrometeorological parameters, Jl. of Agrometeorol. 4: 1-8.

Venkataraman, S. and Kazi, S.K. 1979. A climatic disease calendar for "Tikka" of groundnut. Jl. Maharashtra Agric. Univ. 3: 91-94.

Venkataraman, S. and Krishnan, A. 1992. Crops and Weather. Pub. ICAR. 586-591 pp.

Waggoner, P.E. 1960. Forecasting epidemics in "Plant Pathology" Eds. Horsefall and Diamond. Acad, Press. 292-312.

Wallin, J.R. and Hoyman, W.G. 1954. Forecasting late blight in North Dakota. Bull. N. Dakota Agric. Expt. Stn. 16: 226-231.

Wang, Z.J. et. al., 2009. Bt Cotton in China: Are secondary insect infestations offsetting the benefits in farmers' fields? Agricultural Sciences in China, 8: 83-90.

Wellington, W.G. 1954. Weather and climate in forest entomology. Met. Monog. Am. Met. Soc. 2:   11-18.

Whitman, J.A., Anders, M.M.; Row, V.R. and Reddy, M. 1995. Management of Helicoverpa armigera (Lepidoptera noctuidae) on chickpea in South India; Thresholds and economics of host plant resistance and insecticide application. Crop Protection, 4: 37-46.

Wili, G.M. 1985. Comparison of the design of two volumetric spore traps. Phytopathol. 75: 380 pp.

Wilson, L. T. and W. W. Barnett. 1983. Degree-Days: An Aid in Crop and Pest Management. California Agriculture. 37: 4-7.

Xu, X. M. 1996. Effects of temperature on latent period of rose powdery mildew, Podosphaera pannosa. Plant Pathol. 48: 662-667.

Yella Reddy, D. et. al., 2006. Dynamic cumulative weather based index for forewarning of rice blast. Jl of. Agrometeorol. 8: 1-6.

Yoshimura, S. and Tagami, Y. 1967. Forecasting and control of bacterial leaf blight of rice in Japan, Proc. Symp. Tropical Agric. Res. 25-38 pp.

Young, W.W. 1966. Cited by De Villiers, 1966.

Zahid, M.A. et. al., 2008. Determination of economic injury levels of Helicoverpa armigera (Hubner) on Chickpea. Bangladesh Agric. Res. 33: 555-563.

Zalom, F.G. et. al., 1990. Degree-days: the calculation and use of heat units in pest management. Div. Agric. and Natural Resources. Univ. of Califiornia, Davis, USA . 10 pp.

Zalom, F.G. et. al., 1990. Monitoring tomato fruit worm eggs in processing tomatoes. Calif. Agric. 44: 12-15.

CHAPTER 8

Erosion of Top Soil by Rain and Wind

It is possible to raise plants using inert material like vermiculite, a by product from the mica mines. Plant culture is also possible through use of nutrient solutions. The latter is called hydroponics. However, soil-less culture is confined to laboratory use or in raising high value off-season produce or as a hobby. For the purpose of feeding populace, soil culture is a must. Soil is formed by the disintegration of parent rock material by natural, physical and chemical processes. In each of the above processes, weather parameters play a dominant role. Therefore, the process of formation of soils is termed weathering. The role of weather in soil formation is recognised by soil scientists in the classification of soil types. The effect of climate on soil formation results in different soil types being formed from same parent material in different regions and same soil type from different parent materials evolving at a given place (Venkataraman and Krishnan, 1992).

Fully weathered parent rock material is present as soil in the top layers. The top soil used in agriculture is the resultant of climatic action extending over may centuries. The rate of formation of top soil is much lower than the accumulation of fossil fuels. Alternative, renewable energy sources can be located and used to cope with dwindling fossil fuel supplies, No such strategy is possible in case of soils. Thus, top soil is a very valuable resource that cannot be allowed to be depleted quantitatively. Removal of soil by wind and rain called wind-erosion and rainfall-erosion respectively do occur in nature. Wind and rain are outside human control. Soil erosion is accelerated by anthropogenic actions like deforestation and bad land and water management practices in crop culture which lead to detachment of small soil particles from the surface resulting in local, everyday soil erosion. Soil erosion by rain and wind can and needs to be mitigated through control measures. For this, a proper understanding of the physics of the processes of wind and rainfall erosion and liability of a place to erosion by wind and rain are necessary.

8.1 Physical Explanation of Soil Erosion by Rainfall

Rainfall brings about soil erosion through its impact effect on soil surface and surface run off. Any theory of rainfall action on soil erosion must take account of the following facts observed by many workers, namely: (i) lack of association between soil loss and rainfall amounts even when individual rain storms of same magnitude are considered (ii) reduction of erosion due to plant cover or a mechanical cover including a thin gauze and (iii) close association between rainfall intensity and soil erosion. We may examine the action of rain from the above angles

8.1.1 Splash Erosion

The role of impact of rain drops in the soil erosion process was recognised as early as 1877 by the German scientist Woolny. Ellison's work (1944, 1945) on the mechanical action of raindrops on soil showed (i) splash erosion as a major factor in water erosion (ii) the futility of erosion control measures only against the scouring effects of rain and (iii) importance of vegetation cover against rainfall erosion.

In the field, splash erosion occurs as detailed below. Soil aggregates in general are sufficiently big to resist easy removal by water. Raindrops, falling on a moist soil aggregate, exert a mechanical effect at the point of impact and loosen the soil particles. When an aggregate is fully wetted, further rain penetrates it by capillarity and compresses the enclosed air. When the compression is sufficiently built up, the aggregate explodes resulting in loose soil particles. Now, the rain drop bursts into small splashing drops on impact. The loosened soil particles then become suspended within the small splashing drops which extract kinetic energy from the primary drop. The splashing water drops land at some distance from the point of initial impact. On a horizontal surface the suspended particles show no preferred direction of displacement in the horizontal. On a slope, the splashing drops land at a distance closer to the initial point of impact upslope than those landing down the slope. Due to this, there is a movement down the slope of the suspended soil particles in the splashing drop and a deposition of soil particles, usually at the foot of the slope, occurs. This type of erosion is called "Splash Erosion"

For splash erosion, raindrops use their kinetic energy, which again depends on the size of the drop and its terminal velocity. The surface tension of water is insufficient to maintain a water drop bigger than 9 mm. In natural conditions, raindrop diameter rarely exceeds 5 mm. The terminal velocity increases with diameter. A rain drop of 4 mm diameter can reach a terminal velocity of about 34 cm/sec and hence have potential energy to lift a sand layer of 1 square centimeter area and 1.33 mm thickness up to a height of about 6 cm. Thus, large drops obtain sufficient energy to disturb the top soil structure on impact.

8.1.2   Rainfall Intensity

Extensive studies have shown that the size distribution of rain drops in a rain is closely associated with rainfall intensity. Defining a Median Diameter, MD as the one in which particles with above MD contribute an equal volume as particles below MD, a relationship was found between MD and rainfall intensity (Best 1950; Ekern, 1953). MD stabilises around an intensity of 100 mm per hour (Hudson, 1963) and tends to decrease slightly at higher intensities due to greater bursting of larger water drops during their fall. Even in tropics rainfall intensity hardly exceeds 150 mm per hour.

From the above it is seen that from rainfall intensity, one can get MD and from MD, the drop size distribution. This together with the known terminal velocities, permits momentum and energy to be computed. Whether raindrop momentum or raindrop energy is to be used to represent the result of impact of falling drops on soil surface is a matter of debate (Rose, 1960).

The curves connecting momentum or kinetic energy to intensity are similar, particularly when intensity exceeds 25 mm per hour, which is the threshold value for commencement of erosive action. The above explains as to why rainfall intensity is the most important property of erosivity of rainfall.

8.1.3   Run-Off

The fine soil particles suspended in the splashing rain drops plug the soil pores, and decrease permeability of top soil. This, along with increase the compression of soil surface, results in a decrease in its infiltration rate. With passage of time, the infiltration rate is reduced below that of rainfall intensity. Water then inundates a flat soil surface or runs off it along a slope. On wetting, these surfaces become more resistant to entry of water and aid run off.

The relationship between surface runoff and rain is influenced by the degree of slope of the soil surface and the amount of water at the time of commencement of action by rain. Computation of surface run off for hydrological purposes is more difficult than in rainfall budgeting studies. In the latter case, surface runoff is treated as one of the non-crop use losses along with evaporation. Percolation and deep drainage. However, surface run off connected with soil erosion can occur even when soil moisture content is below field capacity. However, it is possible to estimate the same through determination of The Antecedent Precipitation Index (API) and Daily rainfall (WMO, 1974) or through calculations by the SCS curve number procedure (USDA, 1972). In fact, the latter technique has been used to compute surface run-off (i) in the construction and management of on-farm-reservoirs for rice cum fish culture in Eastern India (Pandey et al., 2005) and (ii) in watershed areas of (a) Palestine (Al-jabari et al., 2009), Goodwin Greek, USA (King et al., 1999) and (b) Warsagaon Dam Catchment, Maharashtra, India (Kulkarni et al., 2004).

8.1.4   Scouring

When water flows down a slope the rain drops falling on them causes turbulence in the soil layers. This helps soil particles even up to 1 cm diameter to remain suspended and transported. The streaming soil water suspension further loosens the soil particles from the soil. This action, called scouring, grinds out a number of micro-channels called Rills which may successively join one another to form wider and deeper depressions called Gullies. Erosion resulting from scouring of Rills and Gullies are called Rill erosion and Gully erosion respectively. Scouring of flood plains and beds and banks of channels and streams due to energy exerted by force of concentrated flow of water constitutes Sheet Erosion which can result in mass removal and transport of soil material.

8.2   Soil Erosion Models

Soil erosion by rainfall is also influenced by terrain and crop and management practices. Comprehensive erosion models covering all the above aspects are of great value for use in soil conservation projects. Several workers have tried to develop soil erosion models. The models attempted to be evolved can be classed as process-based physical models, empirical models based solely on observed data and conceptual models, The physical models mathematically detail the processes involved in soil erosion, namely, detachment, transport and deposition of soil particles and the solving of equations relating to the above processes provide estimates of soil loss. It has not been possible to develop a fully process-based model and many physical models have to take recourse to use of only observed data in some aspects of model development. The empirical models relate observed soil loss directly to management and environmental factors through statistical relationships and are, therefore, site, crop and management specific and of local applicability. The conceptual models lie somewhere between the physical and empirical ones with emphasis on adaptability rather than adoptability. A detailed discussion relating to the nature of the above three models have been presented by Lane et al., (1988).

8.2.1   Prediction of Soil Loss

8.2.1.1 Universal Soil Loss Equation

The Universal Soil Loss Equation (USLE), an empirically-derived mathematical model has been developed in the USA based on soil erosion data collected since the 1930s in USA. The USLE seeks to estimate the average annual soil loss as the product of a series of independent factors like Rainfall Runoff, Soil Erodibility, Length and Steepness of slope, Crop and Land Management and Conservation Practices like terracing, strip cropping and contouring. In the above the Rainfall-Runoff factor is expressed as a product of kinetic energy in a rainstorm times the rainfall amount and maximum intensity over a given periods. Soil erodibility is expressed in terms relative to a slope of length of 22.6 meters and slope of 9%.

USLE is empirical and site specific and would require measurements over a wide range of sites for its solution, calibration and validation (Babau, 1983). Roose (1977) points out that in USLE the respective weight of terms can vary from 1 to 1000 for Rainfall-Runoff, from 1 to 12 for Soil Erodibility, 1 to 25 for product of slope and length from 1 to 10 for Crop and Management.

8.3   Rainfall Erosive Capacity

As the influence of Rainfall is far greater than that of others, by studying the Rainfall-Runoff factor and charting rainfall erosive capacity one can locate areas most seriously prone to erosion by rainfall.

8.3.1   Computation of Rainfall Energy

In 1978, The FAO/UNEP Expert Consultation on Methodology of Soil Degradation Assessment (FAO/UNEP, 1978) came to adopt the USLE with the stipulation that maximal rainfall intensity over a 30 minute interval be used as a measure for calculation of rainfall energy. Equations for computation of kinetic energy of rainstorms from rainfall intensity have been proposed by many workers. Comparison of rainfall energies computed by the methods of Wischmeier (1959) and Hudson (1971), as given below, are seen to give very close estimates (WMO, 1983).

E in joules of rain and I in mm/hr, $E = 8.7 \log I + 11.9$ (Wischmeier, 1959)

E in joules/m^2/mm of rain and I in mm/hr $= 29.82 - (127.51/I)$ (Hudson, 1971)

8.3.2   Method of Computation of Rainfall Energy, R

An exemplary procedure for determination of Rainfall Energy from autographic rainfall charts has been presented by Babau (1983). For conversion of rainfall intensities into kinetic and total energy, one needs self-recording rainfall charts with a large space of chart per mm of rainfall and per hour. Any errors in recording of rainfall amounts and lack of accuracy in running of the clockwork must be corrected for. The minimum data period for evaluating the Erosion Index, R is 10 years. Data of 22 or more years would be ideal.

The procedural formalities of Babau (1983) for computation of R are as follows:

The rainfall trace must be broken into a number of segments of nearly uniform slopes i.e., intensities. The duration and amount of rain in each segment must be noted.

The value of the kinetic energy (KE) is to be obtained from appropriate relations connecting KE to rainfall intensity I.

The value of KE for each segment is then multiplied by the total rainfall RR in the segment.

The values of KE × RR are added up.

The maximum value of E30 is derived from rainfall charts.

When rain lasts less than 30 minutes the value normally adopted for I30 is twice the total amount of precipitation in the period.

I30 is then multiplied by the total value of KE to get the value of R.

The calculations must be carried out for each spell of rain in which falls are separated by 6 hours in which less than 0.05 inch of rain fell.

8.3.2.1 Periods and Ratings, R

R can be computed in terms of cropping periods, namely: seeding, establishment, growth. Reproduction, maturity and bare soil by summing up the storm values over the period. The advantage of crop period is that within each period, influence of given soil physical conditions and residue effect will be minimal. The ratings for R as determined above are as follows: "Slight" for R values of 0 to 50; "Moderate" for R values of 51 to 500; "High" for R values of 501 to 1000; "Very High" for R values greater than 1000. The rating for the above classes are: 0.5 for Slight, 1.0 for Moderate and 2.0 for High. Regarding soil texture the ratings are 0.2, 0.3, 0.11 and 0.5 respectively for Coarse, Medium. Fine and Stony soils. Regarding Topography the ratings are: 0.35 for slopes of 0 to 8% to, 3.5 for slopes of 9 to 30% and 11.0 for slopes greater than 30%.

8.3.3   Need for basic Determination of Rainfall Energy

The FAO/UNEP Expert consultation on Methodology of Soil Degradation Assessment (FAO/UNEP, 1978) has postulated that Erosivity be calculated by the relationship $R = p^2/P$ where R is Rainfall Erosivity, p is monthly precipitation and P is Total annual precipitation. WMO (1974; 1983) has proposed a number of relationships for indirect estimation of rainfall intensity from more easily determinable parameters. Empirical formula for easy and indirect estimation of Erosivity of rainfall is not advisable. Limited work on rainfall Erosivity shows that erosive capacity of rainfall in tropics will be several times greater than in temperate zones (WMO, 1983). Therefore, there is no alternative in the tropics to calculation of R by Wischmier's method for locations typical of various soil-climate regimes before attempting an indirect estimation of R with easily measured rainfall parameters or maximisation of data-use and for mapping of R for a country as a whole.

8.3.3.1 Extrapolation of Findings

The network density of self-recording rain gauge stations is never dense enough to prepare Iso-Erosivity maps for any given region. Recourse is, therefore, taken to establish relationships between Erosion Indices and other more widely available parameters like rainfall amounts. Annual or seasonal values of Rainfall Erosivity, RE can be sought to be related to annual or seasonal rainfall amounts and the established relations can be used to extrapolate measured values of RE through more widely available rainfall data. Such a procedure has been adopted by Dhyani et al., (2000) to prepare Iso-Erodent maps on an annual basis and for the rainy season for the state of Madhya Pradesh in India.

8.4   Erosion of Soil by Wind

The action of wind on soil leads to the phenomena of detachment of soil particles from the surface, their transport and subsequent deposition. Wind is not uniform either in speed or in direction. The direction of wind can oscillate even in short period of time. There are sudden increases in wind speeds with weaker wind in between, the wind flow is then said to be gusty. Sometimes winds may strengthen up to 20 to 30 meters per second over brief periods and limited areas with change in wind direction. Such bursts are known as squalls. At low wind speeds of 1 to 2 meters per second, the flow is in parallel streams and such a flow is called a "Laminar Flow". At higher speeds wind moves up and down as Eddies and the flow is then said to be "Turbulent". Removal of soil by wind occurs when the force of wind acting on the soil surface becomes strong enough to overcome force of gravity. Such removal of soil by wind constitutes wind erosion.

8.4.1   Erodibility of Soils by Wind

The amount of soil erosion by wind is not solely dependent on wind speed impinging on the surface but also by the erodibility of soils. The soil surface is made up of single particles of various diameters and soil aggregates of various sizes and configuration. The erodibility by wind of a given soil particle depends on its weight. Thus, soil erodibility is governed by the particle size composition and bulk density of the soil surface. Extremely small and light particles of less than 0.1 mm in diameter, heavy particles of greater than 1 mm in diameter and large aggregates greater than 2.5 mm in size are not easily eroded. Now, particles less than 0.1 mm constitute Dust and the mention of resistance of Dust to erosion needs some explanation as wind-carried dust clouds are visibly seen up to great heights. The resistance to wind of Dust particle less than 0.02 mm in diameter is only partly due to their cohesion. Because of their small size, they remain inside the layer of laminar flow of wind and hence are free from the turbulent influence of air flow. Particles greater than 0.05 mm disrupt the laminar flow of wind sufficiently to deflate fine soil particles (Chepil, 1945). Similarly, if the soil is rough with projections of greater than 0.05 mm in height, the cohesion of dust particles is broken. The dust particles then latch on to large deflated soil grains and are removed with them. Once airborne, dust particles can be lifted up to great heights by convective currents.

8.4.2   Forms of Soil Erosion

8.4.2.1   Saltation

When wind impinges on a soil particle, due to the Bernoulli Effect, the static pressure on the top of the soil grain is reduced compared to that at the bottom. This difference in static pressure causes a lift on the grain and the soil particle jumps into free air. This movement of the soil particle is called "Saltation". The saltating particles rise up to 15 to 30 cm and very rarely up to 1 meter. The horizontal length of the jump is about 50% of the height of the jump. The saltating grains dash against other grains, disintegrate them and make them more erodible. They also cause deflection of the tiny dust particle. While saltation is more important in wind erosion than water erosion, it also causes two other types of movement viz Surface Creep and Suspension.

8.4.2.2 Surface Creep

It is the movement, by sliding or rolling along the surface of the ground, of coarse grains, which are too big to be lifted off by saltation. Surface Creep is concerned with particles greater than 0.5 mm and less than 2.0 mm in diameter.

8.4.2.3 Suspension

It is concerned with particles smaller than 0.10 diameters, which are not normally moved by even high wind velocity. Suspension can occur only when turbulent energy of fluid is sufficient to overcome the settling velocity of the particle due to gravitation. Thus, suspension is pronounced under convective conditions when there are up draughts with wind speeds greater than the terminal velocities of particles. These particles continue to float in the higher reaches of the atmosphere even after convection had died down and may be carried away for considerable distances.

8.4.2.4 Mass Drift

Soil erosion not only removes valuable top soil but in arid and semi arid areas induces spread of sand over agricultural and grazing lands, thus rendering fertile lands infertile. This is reflected in the formation of sand dunes and hummocks. Mass drift manifests itself in the migration of sand dunes in desert areas. For this there must be initial areas of removal and deposits, with intervening areas of transport. After sometime, when the initial area of removal is denuded of erodible sand, the initial deposit area becomes one of removal and there is a migration of areas of deposition.

8.4.3 Types of Wind Erosion

The wind erosion of soil is classified as Everyday Soil Erosion, Drifting Dust, Dust Whirls and Blowing Sand.

8.4.3.1 Everyday Soil Erosion

It is a local one and is caused by the wearing action of wind of even low speeds and small gusts. Human activities, like use of farm implements, which lead to detachment of small soil particles from the surface layer, aid everyday soil erosion. Soil degradation proceeds in an unobtrusive but sure manner on a local scale due to everyday soil erosion. During periods of high winds everyday erosion produces potential areas for blowing dust.

8.4.3.2 Drifting Dust

Transfer of soil particles may occur when they are raised into the atmosphere and are deposited around obstacles. Such a process is called Drifting. Very high wind speeds are not required for drifting to occur. If the drifting material is confined to 2 meters, it is called "Drifting Dust", whose areal coverage is small. Drifting dust is associated with convective type of clouds and can occur with speeds of 3 to 5 m/sec. It is extremely difficult to distinguish a strong drifting dust from a weakly blowing dust storm.

8.4.3.3 Dust Whirls

Under strong heating of the surface and low wind speeds, conditions for strong and limited convection are created and Dust Whirls occur. The heights up to which whirls extend depend on the climatic regime and vegetative cover. It increases with aridity and lack of vegetative cover. However, even in deserts, whirls do not reach up to 1 km. They have a diameter of 1/3 to 1/2 of the height. Thus, Dust Whirls cause little soil erosion.

8.4.3.4 Blowing Dust

It may be deemed to occur when dust particles are raised by strong wind into the air causing turbidity of the atmosphere and reduced visibility. Weak, moderate and strong blowing dust may reduce visibility to 2 km, 1 km and less than 1000 meters respectively. Blowing dust is highly frequent in arid regions and over large areas. Besides causing soil degradation over extensive agricultural and grazing areas, Blowing Dust causes high air pollution through increased dust content of air. For blowing dust to occur, soil degradation must have progressed for some time and strong winds with speed exceeding 15 m/sec and bare ploughed up areas with powdered soil are required. Soil erosion is speeded up by Blowing Dust.

8.4.4 Influence of Non-Weather Factors

8.4.4.1 Type of surface

When wind speed is plotted against the logarithm of height, a straight line intercepting the height axis above zero results. This height of interception is known as the mean aerodynamic surface. Wind speed is related to height (Brunt, 1944) as follows:

V_z = 5.75 V_x log(Z/K)/where V_x is known as Drag Velocity, V_z is wind speed at height Z and K is height above the mean aerodynamic surface. K is a measure of aerodynamic surface roughness and is higher for rougher surfaces.

In considering roughness one should distinguish between macro-scale (natural areas) and micro-scale (farm-field) roughness. Open sea is aerodynamically more smooth and sheltered fields and town sites are the most rough. Micro-scale roughness is contributed to by unevenness of soil surface and topographic relief. However, a rough soil surface is not necessarily more erodible than a smooth one, if the roughness is contributed by bigger aggregates and larger particles or a vegetative cover or crop residues.

Relief can have a significant effect on soil erosion. In unobstructed plains, the winds are high and due to higher impact on soil particles, serious erosion results. Unevenness in relief acts in the same way as obstructions and reduce erosion and blowing dust gives way to everyday erosion and more and more soil is removed. Thus, great care must be taken in opening up slopes for cultivation in windy climatic regimes.

8.4.4.2 Soil Properties

A soil is liable for erosion when the ratio of large lumps with diameter greater than 1 mm to that with diameters less than 1 mm becomes 1.0 in the 0 to 5 cm soil layer. The smaller the ratio, the greater is the erodibility to wind. The size of the clay fractions has a great influence on wind erodibility of the soil-the larger the fractions the greater are the resistance to wind erosion.

8.4.4.3 Vegetative Cover and Land use

Removal of vegetative cover and pulverisation of top soil initiate and/or worsen soil erosion by wind. Opening up of natural areas for growing food crops becomes unavoidable many times. A ploughed soil is less resistant to wind erosion than an unploughed one. Land has to be frequently prepared to keep it under crop cover. In either case, soil erosion gets accentuated. Unrestricted grazing by diminution of vegetative cover and breaking up of soil leads to serious soil degradation and erosion.

8.4.4.4 Climate

An increase in continentality, characterised by high range of soil and air temperatures and reduced rain and hydrothermic coefficient, tends to strengthen wind speed. Wind accelerates soil desiccation and exposes soil crevices to erosion by rain. Rain wets the soil surface and enhances soil particle cohesion and leads to enhancement of vegetative cover. Thus, lack of rain during periods of strong winds is important. Freezing and thawing of soil also enhance soil erodibility.

8.4.4.5 Soil Moisture Content

Soil moisture influences erodibility of soils (Chepil, 1956). Moist soil layers have increased resistance to erosion. The level of soil moisture at which erosion can start is the "Critical Moisture Content". For sandy and sandy loams the critical moisture equals the hygroscopic coefficient. For other soils it is the Permanent Wilting Point. Thus, extremely dry soils are required for commencement of erosion. Strong winds can desiccate an even moist soil which is then eroded. The next layer then gets exposed and the process goes on. So what is important is wind speed and its duration.

8.4.5   Deposition of Soil Particles

When wind slows down over depressions, some or all the soil material carried by it gets deposited on the ground. The proportion of large lumps increase on the surface. They collect in small ridges which are oriented in the wind direction. These are known as the Ripples, which move at a speed of about 5% of the speed of the wind. Horizontal whirls are also produced in the depressions. In gorges and ravines, deposits mainly occur on the leeward slopes. When obstructed solidly, the soil particle is deposited in the form of a ridge with the crest of the ridge some distance away from the obstacle with a gentle windward slope immediately in front of the obstacle. With objects permeable to wind flow, fine earth accumulates both leeward and windward of the obstacles. The relative accumulation depends on the extent of permeability of obstacles. Around tree strips, the larger fractions are deposited on the windward side and the finer ones are deposited on the leeward side. The above are of importance in initiation of measures against soil erosion which is discussed in some detail subsequently.

8.4.6   Estimation of Risk and Amount of Soil Loss by Wind Erosion

Wind flow in nature is always turbulent. In a truly turbulent wind field, the wind speed at height Z may be expressed by the following relationship, namely:

$$U(z) = 5.75 \sqrt{T/P} . \log Z/Z_0 \qquad \qquad .....(8.1)$$

Where T is Drag per Unit Area exerted by wind at ground surface; P is Air density; $U(z)$ is mean speed at height z and Z_0 is a constant such that $U(z_0)$ is nearly equal to 0.

The quantity Z_0 is called the Roughness Length. The value of Z_0 for vegetation lies below the maximum height of the vegetation and is essentially a function of the vegetation. This upward displacement by vegetation of the height, above which velocity gradient postulations of wind apply, is called the Zero Plane Displacement and is denoted by "d". Therefore, for vegetation Equation 1 must be rewritten as follows, namely:

$$U(z) = 5.75 \sqrt{T/P} . \log (z-d/z_0) \qquad \qquad .....(8.2)$$

Force exerted by a moving fluid on a submerged body is proportional to the square of the velocity of the fluid. We may, thus write T/P in equations 1 and 2 as equal to U_x so that $T = P.U_x 2$ and U_x is called the frictional velocity associated with and proportional to the drag exerted on the surface by the moving stream.

8.4.6.1 Critical Wind Speed

Till friction velocity U_x becomes sufficient to lift up soil particles, the surface remains undisturbed. As V_x increases, more and more particles are dislodged. The minimal or threshold velocity of V_x at which particles are removed depends on density of the particles. Once erosion has started, the impact velocity required to sustain erosion is lesser than the threshold due to energy extraction from wind by the saltating particles. The difference between threshold and impact velocity is very small for the most erodible soil particles. Under field conditions, soil becomes more erodible with each wind-storm. The soils often have a thin crust. To break through this, a higher velocity is needed. If the crust is moist, the high wind must flow for a longer time. Thus, there can be a range of threshold values ranging from that for a sand dune to that of a non-eroded soil. V_x can be obtained as a product of diameter and specific gravity for various soil particles.

The minimal value of V_x at surface to initiate erosion in terms of wind speed at height of one foot over quite a range of surface conditions have been given by Chepil (1945) and Chepil and Woodruff (1963). Tables to convert wind speed at a given height to the desired one at foot level are available. The above information should be of help in assessing wind erosion risk from climatological wind-speed frequency data.

8.4.7   Wind Erosion Equation

Like the USLE a Wind Erosion Equation had been presented by Woodruff and Siddaway, (1965). However, unlike the USLE, in the wind erosion equation, simple multiplication of the value of the factors is not possible because of the interactions between the several factors. Considering the data requirements, the wind erosion equation is too detailed to be solved. Chepil and woodruff (1963) have drawn attention to the research needed for practical application of the wind erosion equation.

8.4.7.1 The Wind Erosion Index

As mentioned earlier, critical wind speeds required to initiate soil erosion have been given by several workers for different roughnesses of soil surfaces and various classes of soil particles. Bagnold (1953) found the rate of erosion to increase with the cube of the wind speed above a given threshold level. Formulae for computation of the Wind Erodibility Index, "C" has been given by many workers. The one proposed by Chepil and Woodruff (1963) but modified and accepted by FAO (1979) is of the form:

$$\sum C = 1/100 \, V_3 \, ((PET\text{-}P)/PET) * n$$

Where V is mean monthly wind speed at 2 meters height in m/sec.

> P I is Rainfall in the month in mm
> PET is monthly Potential Evapotraspiration in mm and
> n = number of days in the month
> C is the monthly wind Erosivity Index
> [(PET-P) PET] *n gives the number of erosive days in a month

and $\sum C$ is the cumulative values of C for the 12 months of the year.

In irrigated areas the quantity of water supplied by irrigation, say Q is added to monthly rainfall in computing C. As topography has a direct effect on wind speed, this factor is not taken into account for computing the Wind Erosivity Index.

8.4.7.2 Ratings

The ratings for wind erosivity in terms of Values of C as given by FAO (1979) are: Nil to light for 0 to 20; moderate for 21 to 50; Strong for 51 to 150 and very strong for C values > 150.

8.4.8 Wind Erosion Measures

Increasing the lumpiness of the surface will reduce erosion but the effect will be temporary as stability of percentage distribution of clods of various given diameters is weather-dependent. Applying organic manure reduces wind erosion by improving structure and physical properties of soils and by bushy development of plants. Keeping post-harvest stubble improves the roughness of the soil surface and reduces erosivity of wind at the surface. However, having a continuous crop cover is often not possible in many areas and even periods.

8.4.8.1 Wind Barriers and Shelter Belts

The most important and the most effective measure against wind erosion is the use of wind-breaks, wind barriers and shelter belts, which act as an impediment to wind flow. Strong winds generally follow a contour. An obstruction induces an on-coming wind to break away and separate from the surface. Below the level of separation, a partially protected zone of reduced overall mean wind speed is established. In this zone the risk of soil erosion by wind is reduced.

Within the realms of wind speeds encountered in nature and the area of field protection desired, the percentage reduction in wind speed downwind is independent of wind speed for a given obstruction and field. However, even for the same period of wind, obstruction and field, the area of protection will vary with the angle of incidence of the wind with respect to the obstruction.

The efficiency of windbreaks depends on their permeability and dimensions. A solid obstruction is not suitable as a windbreak. A

Dense barrier can protect an area 15 to 20 times its height. By increasing its permeability to 50%, the area of protection would increase to 20 to 25 times the height of the barrier. Dense barriers provide large wind reductions over a short distance while porous barriers provide lesser wind reduction but over greater distances. For best wind reduction, shelters more porous in the lower heights, logarithmic increase in porosity with height and with total gaps in the barrier equal to about 30 to 50% area of the belt would be ideal. The gaps should not be too large to prevent scouring by air of soil immediately leeward of the shelter. For best results, the barrier must be perpendicular to the wind flow.

There is no cumulative effect of shelter belts. The effectiveness is the same whether there is a single belt or it is within a staggered system of parallel belts. Trees as wind breaks are not suitable as they take up much cultivable space and compete with crops for nutrients and water. Orientation of the rows such that the wind strikes them perpendicularly would be very effective, Strip cropping and crop barriers are quite useful. Effectiveness of strip cropping could be increased by deciding the width of strips in relation to the extent of deviation of erosive winds from perpendicularity to the crop rows (Chepil and Woodruff, 1963) - greater the deviation the narrower should be the strip.

8.4.9   Predominant Wind Erosion Direction

For effectively reducing wind erosion, properly distanced wind breaks of appropriate height must be oriented perpendicular to the dominant wind direction. For this, determination of the predominant wind direction and wind speeds associated with it are required to be known. Climatically the wind data are depicted as "Wind Roses" in which the frequency of calm is indicated in the centre of a circle and the frequency of occurrence of wind from one direction is indicated by a line of length proportional to the frequency. The frequency of occurrence of given ranges of wind speed for each of the directions of the wind is also given. Wind Roses are prepared on a monthly basis and one can get the monthly march of the dominant wind direction. However, for soil erosion control, for each direction the summation of the product of the percentage time the wind flows from that direction and the cube of the wind speed associated with it is taken. Now a line is drawn towards the centre of the wind rose from the 16 main directions with a length proportional to the summed up value. Since wind erosion in a field can be deemed to be the same for wind blowing from one direction or from the opposite direction, the joint length of a pair of diametrically opposite lines is taken and is divided by the total length of all the lines. When this ratio exceeds 25% the direction is deemed to be the significantly prevailing wind erosion direction (Chepil et al., 1962).

References

Al-Jabari, S.; Sharkh, M.A. and Al-Mimi, Z. 2009. Estimation of runoff from agricultural watershed using SCS curve number and GIS. Proc, Thirteenth International Water Technology conference, IWTC 13. Hurghada, Egypt. 1213-1229.

Babau, M.C. 1983. The erosive capacity of rainfall, Pub. World Met. Org. WMO, No. WCP 41. 28 pp.

Bagnold, R.A. 1953. Surface movement of blown sand in relation to Meteorology. Proc. International Symp. On Desert Research. 89-96 pp.

Best, A.C. 1950. The size distribution of rain drops. Quart, Jl. Roy. Met. Soc. 76: 16-36.

Brunt, D. 1944. Physical and Dynamical Meteorology. Cambridge Univ. Press. London, 428 pp.

Chepil, W.S. 1945. Dynamics of wind erosion. II. Initiation of soil movement by wind. Soil Sci. 60: 397-411.

Chepil, W.S. 1956. Influence of moisture on erodibility of soils by wind. Proc. Amer. Soc. Soil Sci. 20: 288-292.

Chepil, W.S. and Woodruff, N.P. 1963. The physics of wind erosion and its control. Advances in Agronomy. 15: 211-302.

Chepil. W.S.; Siddaway, F.H. and Armbrust, D.V. 1962. Climatic factor for estimating the wind erodibility of farm fields, Jl. Soil Water Conservation. 9: 67-70.

Dhyani, M.L, Kumar N.; Tandon, R. and Babu, R. 2000. Rainfall erosion potential and Iso-erodent map of Madhya Pradesh. Jl. of Agrometeorol. 2: 103-112.

Ellison, W.D. 1944. Studies of raindrop erosion. Agric. Eng. 25: 131-136 and 181-182.

Ellison, W.D. 1945. Some effects of rain drops and surface flow on soil erosion and infiltration. Trans. Amer. Geophys. Uni. 26: 415-429.

Ekern, P.C. 1953. Problems of raindrop impact erosion. Agricl. Eng. 34: 23-25.

FAO (1979). A provisional methodology for soil degradation assessment. FAO, Rome, 88 pp.

FAO/UNEP (1978). Report of an expert consultation on methodology for assessing soil degradation. FAO, Rome. 70 pp.

Hudson, N. 1963. Rainfall size distribution in high intensity storm. Rhod. Jl. Agricl . Res, 1: 6-11.

Hudson, N. 1971. Soil Conservation. Batsford London, 330 pp.

King, K.W.; Arnold, K.G. and Bingner, R.L. 1999. Comparison of Green-AMPF and curve number methods on Goodwin creek watershed using SWAT. Trans. Amer. Soc. Agricl. Engineers 4: 919-925.

Kulkarni, A.A.; Agarwal, P. and Das, K.K. 2004. Estimation of surface rainfall-runoff modeling of Warsgaon dam catchment- A geo-spatial approach. Pub. Ind. Inst. Remote Sensing, Dehradun, India, 18 pp.

Lane, L.J.; Shirley, E.D. and Singh, V.P. 1988. Modeling erosion on hill slopes. In: Modelling Geomorphological Systems. John Wiley Pub. New York. 287-305 pp.

Pandey, P.K.; Panda, S.N. and Pholane, L.R. 2005. Modeling for maximizing precipitation utilisation in rainfed agriculture in Eastern India. Bull. National Inst. of Ecol. 16: 113-120.

Roose, E.J. 1977. Erosion et unissellement en Afrique de l'ouest – uningt. Annes de measures en petites parcelles experimentales Travaux et documents No. 78. Orstom, Paris. 108 pp.

Rose, W. H. 1960. Soil detachment caused by rainfall. Soil Sci. 89: 28-35.

U. S. Dept. Agri. USDA 1972. USDA Soil Conservation Service. National Engineering Handbook. Sec.4. Chapters 4-10. 544 pp.

Venkataraman, S. and Krishnan, A. 1992. Crops and Weather. Pub. Ind. Counc. Agricl. Res. ICAR. 586 pp.

Wischmeier, W.H. 1959. A rainfall erosion index for a universal soil loss equation. Proc. Soil. Sci. Soc. Amer. 23: 246-249.

Woodruff, N. and Siddaway, F.H. 1965. A wind erosion equation. Proc. Soil Sci. Soc. Amer. 29: 602-608.

WMO, 1974. Guide to Hydrological Practices. Pub. WMO, Geneva. No. 168.

WMO, 1983. Meteorological aspects of certain processes affecting soil degradation especially erosion. WMO Technical Note No. 178. 149 pp.

CHAPTER 9

Climate Change

Climate Change is a term very much in vogue now a days. However, the phenomenon of climate change is present in the very rotation of the earth around itself with an inclined axis. With each rotation the position of the axis of the earth shifts infinitesimally and imperceptibly. Accumulated over many centuries, such a shift can lead to reversal of seasons locally. The shift of local seasonal features with time has been noticed in ancient times. Unable to carry out the mathematical calculations required to gauge the shift, the ancients believed that the seasonal features would revert back and termed the phenomenon as "The Oscillation of Equinoxes". It was Newton who showed that the observed changes in seasonal changes were unidirectional and termed it "The Procession of Equinoxes". It is estimated that a full reversal of season from Winter to Summer and vice versa will occur locally once in 45000 years plus or minus 5000 years. There are reports (Bose, 2005) of the Tropics of Cancer shifting southwards at about 1.27 km per 100 years. Due to normal shifts in Earth's obliquity and external gravitational influences, the above will lead to shrinking of the tropics and setting in of harsher climates in the current tropical areas. Again, for some unknown reason the earth gets covered once every one lakh years by ice and this gives rise to inter-glacial periods of warm and ice ages.

The above climatic change occurs on a geological time scale and is of academic interest only. However, the progressive increase in aridity of the once green areas of Harappan and Mohenjodaro civilisations, attributed to reduction in solar radiation due to increased dustiness of the atmosphere, shows that highly perceptible changes in climate can occur locally even over a few centuries. Anthropogenic activities have brought, within the realm of reality, the possibility of perceptible climate changes even within decades.

218

9.1   IPCC

The United Nations Environment Program4me (UNEP) and the World Meteorological Organisation (WMO), established in 1988 the Inter Governmental Panel on Climate Change (IPCC) to periodically assess the state of global environment and advise Members through UN agencies. Since then the subjects of "Global warming" and "Climate Change" have drawn international attention.

9.2   The Greenhouse Effect

Global warming arises due to the emission into the atmosphere of large quantities of Carbon Dioxide, CO_2; Methane, CH_4 and Nitrous Oxide, N_2O on account of industrial emissions, large scale use of fossil fuels and burning of biomass. By interfering with the energy balance of the earth-atmosphere system. Carbon Dioxide and Nitrous Oxide, which are long-lived in the atmosphere, cause global warming. This is similar to what happens in a greenhouse. Thus, the gases Carbon Dioxide, Methane and Nitrous Oxide are referred to as the Green House Gases, GHGs and global warming is termed as the "Greenhouse Effect". Since global warming can, per se, affect many biotic, hydrological, atmospheric and oceanic processes, changes in climate on a global scale can be expected to follow.

9.3   Influences of Climate Change

Life-duration, developmental rhythm, yield, water needs, and fertiliser requirements of crops and incidence of pests, diseases and weeds are influenced by climatic factors. Additionally, all forms of precipitation, carbon dioxide and solar radiation constitute inputs for crop production. Thus, a brief review of influence of climate change on rainfall, greenhouse gases, solar radiation and temperature appears called for and is detailed below.

9.3.1   Rainfall

Rainfall, associated with well organised and annually recurring weather systems, is directly or indirectly affected by oceanic phenomena like El Nino and La Nina and atmospheric phenomenon of Southern Oscillation (SO). The appearance of abnormal warm and cold water off the Peru coast in Eastern Pacific, in the region 14 degrees North to 10 degrees South and 148 to 100 degrees west longitude, are respectively called El Nino and La Nina. The warm waters associated with El Nino move eastwards and warms the atmosphere. Sea level falls and rises respectively in the West and East by 25 centimeters. Reduced upwelling in the Eastern Pacific strengthens the temperature anomaly. El Nino affects normal development of the Indian Summer Monsoon Rainfall (ISMR) by creating a low pressure system over equatorial South Pacific and disturbing the low pressure area over the Indian Ocean that is conducive to good ISMR (Sikka, 1980). By the same analogy, La Nina can lead to bountiful ISMR. El Nino causes severe droughts over Australia and Indonesia and increases Pacific tropical cyclones. Under El Nino, North West and Greater Plains of USA become wetter while North East USA becomes drier. El Nino has become important since the beginning of the 20th century with a frequency of 3 to 7 years. La Nina follows El Nino about one year later. Both El Nino and La Nina events appear to be predictable satisfactorily, 6 to 12 months in advance. The atmospheric phenomenon, SO is a measure of the difference in the atmospheric pressure between the Indian Ocean region and the Pacific region (De 2000). The search for an entity that could give a measure of the large scale ocean-atmosphere interaction processes involved in rain formation led to the concept of ENSO (Rasmusson and Carpenter, 1982). Some of the most pronounced year to year variability in the climatic features in many parts of Asia has been linked to ENSO (IPCC, 2001). In view of the above a review of work done on the influence of global warming on El Nino, La Nina and SO affecting Indian Summer Monsoon Rains (ISMR) appears warranted and is detailed below.

9.3.1.1 Indian Summer Monsoon Rainfall (ISMR)

The analyses of rainfall data in India for the past 120 years (Gadgil and Mishra, 1995; Kripalani and Kulkarni, 1997) indicate that (i) only 7% of La Nina years had drought (ii) no El Nino year was a flood year and (iii) a non El Nino year was likely to be normal. Further studies have shown the complex nature (Khole and De, 1999) and a weakening of the relationship (Ashrit et al., 2001) between ISMR and El Nino events. The lack of an explicit relationship between El Nino and ISMR may be due to (a) non-occurrence of even overlapping periods between the active phases of the two phenomenon (Asnani, 2001) (b) competition for convergence of moist air between the atmosphere over the Pacific ocean and that over the Indian ocean

(Gadgil et al., 2002) (c) need for sea surface warming starting in the Eastern Equatorial Pacific (Mooley and Paolino, 1989) and (d) suppression of the El Nino effect during epochs of above normal rainfall (Kripalani amd Kulkarni, 1997).

Occurrence of El Nino with negative phase of SO is deemed conducive to poor ISMR. Considerable work has been done to analyse the teleconnection between ISMR and ENSO which varies on a decadal time scale (Nigam, 2003). The years 1990 to 1995 were of prolonged warm ENSO while 1997 was the strongest year of ENSO. No large scale drought occurred in the above years and rainfall was in fact abundant in 1994 (Khole and De, 1999; Ashrit et al., 2001). The breakdown in the inverse relationship between ENSO and ISMR in a historically unprecedented way in the recent decades (Krishnakumar et al., 1999b) may be due to Eurasian warming in spring and winter on account of global warming enhancing the land-ocean gradient and hence of a stronger monsoon (Krishnakumar et al., 1999a) with an associated northwards shift and greater spatial variability (Bhaskaran, et al., 1995; Meehl and Washington, 1993. Tett, 1995) and enhanced inter-annual variability (Lal et al., 1995). The likelihood of ISMR entering an epoch of deficient rainfall in the coming years with no reduction in El Nino activity is a worrisome feature. However, due to global warming being beneficial in offsetting magnitude of monsoon failures, the envisaged climate change is likely to lead to less frequent failure but greater variability of ISMR in the near future.

9.3.1.2 Global Rainfall

Individual workers have used various General Circulation Models, GCMs and numerical models of the atmosphere in assessment of increase in green house gases on output of GCMs (Manabe and Wetherald, 1975; Manabe et al., 1992; Washington and Daggupaty, 1975). The aim is to project the magnitude and/or direction of changes of rainfall in different parts of the world due to climate change. Now, there is no simulation model available that can predict either the temporal or spatial variations in rainfall just ahead of the rainy season. The projections not only lack unanimity but also uniformity, particularly on a regional basis (Mitchell et al, 1990).

The IPCC (2001) has made a compilation of the projections made by many workers. These have been adapted by many workers for national projections like Lal et al., (2001) for India. An examination of the agreement of the projections amongst five or more of the 10 models on a regional basis (Neelin et al., 2006) shows fairly good inter-model agreement in the higher latitudes but not in the low latitude tropics where the differences are not due to trends or quantum of changes but in variations in identification of regions of significant changes (Neelin 2009). In light of the above, in the following paragraphs, important and generally acceptable features that emerge from an examination and reconciliation of the reported findings are presented.

El Nino episodes have been predominant since the mid-seventies in the context of the last 100 years (IPCC, 1995). The present frequency and amplitude of ENSO and its precursors is likely to continue in the ensuing years. An increased frequency of ENSO events and shift in their maximal occurrence from January to September in a warmer atmosphere is indicated (Collins, 1999). In future, seasonal precipitation extremes associated with ENSO events are likely to become more. The intensity of extreme rainfall events is projected to be higher in a warmer atmosphere (Lal et al., 2000). An increase in green house gases diminish the impact of El Nino events (Ashrit et al 2001). Increased precipitation in the polar and near polar regions and decreased precipitation for the middle latitudes of both hemispheres are projected as a result of the expected pole ward shift in the jet streams. Increased precipitation in the near equatorial region and a decrease in rainfall in the subtropics are predicted due to a forecast of strengthening of the tropical Hadley cell pattern of atmospheric circulation. In the high latitudes increase in precipitation will occur as rainfall at expense of solid precipitation. The contention that wet areas becoming wetter and dry areas becoming drier (Meehl and Washington, 1996) is based on the assumption of uniform increase in sea surface temperatures everywhere. Although ocean surface temperatures can be expected to increase mostly everywhere by the middle of the century, the increase may differ by up to 1.5 °C depending upon the region. Regions of peak sea surface temperature will get wetter and those relatively cool will get drier. Two patterns stand out. First, the maximum temperature rise in the Pacific is along a broad band at the equator. By anchoring a rain band similar to that during an El Nino, it influences climate around the world through atmospheric teleconnections. A second ocean warming pattern with major impact on rainfall occurs in the Indian Ocean with warming in western India reaching 1.5 °C, while the eastern Indian Ocean it is dampened to around 0.5 °C. Droughts could then beset Indonesia and Australia, whereas regions of India and regions of Africa bordering the Arabian Sea could get more rain. A long-term upward trend of rainfall in tropical areas over both land and sea with higher falls over sea than land has been noticed.

9.3.2   Carbon Dioxide

Evidence from polar ice core samples (Neftal et al., 1985) show that the Carbon Dioxide level in the air has over two centuries increased from 280 ppm in 1773 to 330 parts per million in 1973. However, the next increase of 50 ppm has taken place in only 33 years. Quantum of emission of CO_2 has been estimated as 1.2 ppm (Conway, 1988) and 1.5 ppm (Siegenthaler, 1990). Published data of carbon dioxide concentrations in air show that CO_2 has increased at 1 ppm/year from 1960 to 1970 and this increase had been maintained at 1.5 ppm from 1970 to 2000. In the decade 2001-2010, Carbon dioxide emission had been 2 ppm/year. Currently CO_2 emission is estimated to be occurring at 3 ppm per year. At this rate, the CO_2 concentration by 2050 will be 510 ppm. The normal value of CO_2 concentration assumed in agrometeorological studies is 330 ppm. So by middle of this century CO_2 concentration would have increased by about 50%.

Vegetation is both a sink and source for CO_2 and may be carbon-neutral on an average. Decay of organic matter, industrial emissions and burning of biomass also release CO_2 into the air. However, the major cause for sharp increase in CO_2 emissions is due to the burning of fossil fuels, which injects into the air CO_2 equivalent to 6 ppm. Forests cover 1/3 of the land area and contribute to 2/3 of global photosynthesis. Thus, forests constitute a major terrestrial sink for CO_2 (Lal and Singh, 2000), as important as that of the oceans. Deforestation contributes in a negative way to build up of CO_2 in air to the extent of 25% of that of fossil fuels (Krishna and Saha, 2008). For CO_2 concentration to double the 2000 value of 370 ppm in 2100, as envisaged by IPCC (1990), the average annual emission level has to increase by 50% compared to the current levels. An increase in 25% of fossil fuel consumption or a decrease in the capacity of terrestrial and oceanic sinks for Carbon Dioxide can lead to the above scenario. Considering the history of CO_2 emissions and absorptions from the beginning of this century the above possibility cannot be termed unlikely.

9.3.3   Methane

The concentration of Methane which was 700 parts per billion in 1750 had increased to 1750 ppm or 1.75 ppm in 1990. This is a 250% rise compared to a 25% rise in Carbon Dioxide. Production and transport of coal, natural gas and oil and decay of organic matter in municipal solid waste landfills, rice paddies and ruminating and belching livestock are the major sources for emission of Methane. Rice paddies contribute 25% of the total global production of Methane (Liesach et al., 2002) while the contribution from livestock is 16%.

The identification of wetland rice as a major source of atmospheric Methane has come about a little late as comprehensive measurements of Methane fluxes from flooded paddies had started only in the early 1980s (Cicerone and Shetter, 1981; Cicerone et al., 1983). Puddling of rice fields leads to cutoff of oxygen supply from the atmosphere to the soil and results in anaerobic fermentation of organic matter by methanogenic bacteria and in production of Methane, by trans-methylation of acetic acid and reduction of CO_2. (Neue and Scharpenseel, 1984). Methane remains trapped in an undisturbed, flooded paddy field with only a small amount escaping into air as bubbles Cultural practices associated with irrigated lowland rice account for 30% of the soil methane released into the air while aerenchyma cells of the rice plant provide the conduit for 70% of the Methane released into the air. Methane constitutes hardly 2 ppm of air compared to 350 ppm of CO_2. However, a Methane molecule is 30 times more efficient in trapping of heat compared to a molecule of CO_2. Unlike carbon dioxide, which can remain in the atmosphere for over 100 years, Methane has a half-life of only 8 to 12 years in the atmosphere. It has been reported (Reddy et al., 2005) that higher biomass production of rice and higher incorporation of organic matter to puddle fields increase Methane emissions.

Emission of Methane had slowed down in the 1990s to remain at a steady at a value of 1.75 ppm from 1999 to 2002 (Dlugokencky, et al., 2003) and had increased by only 4% from 1990 to 2000 (Elizabeth and Dina, 2006). By 2020 the concentration of Methane is expected to increase by 40% (Elizabeth and Dina, 2006). Switch over from flooded paddies to the SRI system of rice cultivation can reduce Methane emissions (Roger and Ladha, 1992). Even medium strong measures can reduce Methane emission by 1/8 th (Garg et al., 2004). Therefore, from 1990 to 2020, Methane emissions will increase by 35% and from

2020 to 2050 by a further 53%. Thus, concentration of Methane in air in 2050 will be 3.6 ppm.

9.3.4   Nitrous Oxide

The concentration of Nitrous oxide has steadily increased from a value of 270 ppm in 1750 to 314 ppm by 1998 (IPCC, 2001). 90% of Nitrous Oxide in the atmosphere is of agricultural origin. Because of this, the concentration of Nitrous Oxide is expected to increase by 17% only up to 1920. (Elizabeth and Dina, 2006). Even medium strong measures can reduce Nitrous Oxide emissions by 15% (Garg et al., 2004). So, increase in Nitrous Oxide up to 2020 will be 15% and increase from 2020 to 2050 will be 22.5%. Thus, concentration of Nitrous Oxide in air in 2050 will be 0.330 ppm. Now, a Nitrous Oxide molecule is 270 times more effective trapping heat than a Carbon Dioxide molecule. Like carbon Dioxide, Nitrous oxide can remain in the air for about 100 years.

9.3.5   Cloudiness

Almost all General Circulation Models (GCMs), predict increased evaporation and precipitation due to surface warming and enhanced long-wave radiational cooling from a warmer atmosphere on account of ingestion of GHGs (Mitchell et al., 1987). Increased water vapour feedback amplifies the direct radiative effect of Carbon Dioxide (Pal et al., 2001). Carbon Dioxide absorbs energy in some small segments of the thermal infrared segment that water vapour misses and warms air a bit more. Doubling of Carbon Dioxide is held to be equivalent to a 13% increase in water vapour. Increase in temperature increases the capacity of air to hold water in vapour form. However, continued increase in water vapour will lead to formation of clouds which will reduce radiation receipt and exert a cooling effect. Which of the two processes will eventually dominate and when and where are difficult to predict. Increase in cloudiness associated with increase in precipitation in the rainy or summer season will have negligible effect on crop productivity. What one has to look for is an increase in cloudiness in the normally clear season due to increased carbon dioxide or aerosol concentrations. Doubling of Carbon Dioxide is expected to lead to an increase in mid and high altitude cloudiness (Yao and Del, 1999) while thinning of low clouds in sub-tropics and mid-latitudes with lower albedos due to warming has been indicated (Del and Wolf, 2000). Pal et al., (2001) opine that an increase in Carbon

Dioxide will lead to a reduction in the quantum of outgoing long-wave radiation and result in greater cloudiness with the increase in cloudiness in winter being double that in the summer and rainy season. However, none of the climate models is able to even reasonably replicate the trends in latitudinal variations in mean annual cloudiness. So, one has to take recourse to studies on trends noticed in, quantitative terms, of observed solar radiation or cloudiness in assessing influence of climate change on cloudiness.

9.3.6   Solar Radiation

Theoretically, enhanced Carbon Dioxide should lead to a decrease in mean daily global solar radiation (Zaitao et al., 2004). Reduction in solar radiation at 2% per decade at many stations in India during the clear weather period of May to December has been reported (Shende and Chivate, 2000). Stanhill and Cohen (2001) report that during the past 50 years in many industrial regions of the globe solar radiation has decreased by 2.7% per decade. The reduction in water requirements of winter wheat and summer corn at the rate of 5% and 7.5% per decade respectively in Hebei province China during the period 1965 to 1999 is attributed to a reduction in hours of bright sunshine (Li et al., 2008). The increase in cloudiness is attributed to an increase in anthropogenic aerosols and pollutants changing the optical properties of the atmosphere (Stanhill and Cohen, 2001). Reduction in solar radiation normally equals 2/3rds of the increase in cloudiness. Thus, in a business-as-usual scenario, cloudiness can be expected to increase by 20% by 2050. Quantum of decrease in solar radiation assumed for studies on influence of climate change on productivity of crops are 1, 2 and 3 MJ/m²/day. The above will correspond to 5, 10 and 15% reductions in solar radiation. Solar brightening due to pollution control measures has been noted. So cloudiness can be expected to increase to 15% leading to a reduction in solar radiation by 10% by 2050.

9.3.7   Temperature

Global mean annual temperature has increased by 0.6 °C from 1901 to 2000. IPCC through its assessment reports has been trying to gauge the likely magnitude of mean annual global temperatures in 2050 and 2100 due to global warming relative to that at 2000. An increase of 0.23 °C per decade for the 21st century had been postulated by IPCC in 1996. However, the period 1995-2006 has seen the twelve warmest years since 1850. This is reflected in the differences in assessment of mean global temperatures by the 3rd and 4th assessment reports of IPCC, which shows that temperatures may be increasing by about 0.35 °C per decade from the start of the 21st century.

Assessments of increase in temperatures are based on outputs from General Circulation Models (GCMs) that determine the atmospheric Carbon Dioxide concentration and the temperature increase thereof. Cox et al., (2000) have drawn attention to (i) the accelerating effect of the feedback between climate and the biosphere on climate change (ii) non-inclusion of carbon cycle feedback in many GCMs and (iii) increase in temperature by one degree centigrade by 2100 due to carbon cycle feedback. Many models do not take account of the cooling effects of Sulphate aerosols (Lal et al., 1995; Tett. 1995; Lal and Singh 2001), amounting to one degree centigrade for a doubled CO_2 level (Bhaskaran et al., 1995; Kumar and Ashrit, 2001).

The emphasis is thus on determination of the time of attainment of a doubling in the concentration of Carbon Dioxide which will lead to an increase of temperature of four plus or minus one degree centigrade (Gigori et al., 1989; Schneider, 1989). Allowing for cooling effects of aerosols the increase in temperature in the time interval for doubling of CO_2 levels can be taken as 3 degrees centigrade (McCarthy et al., 2010). IPCC has formulated that the increase in air temperature will vary according to the likely prevalent scenario which includes factors like population density, rate of economic growth and ratio of use of fossil and non-fossil energy sources. Relative to 2000, the rise in temperature by 2100 is projected by IPCC to be about (i) 4 degrees centigrade for a highly pessimistic scenario (ii) 2.5 degrees centigrade for a medium emission and business as usual scenarios and (iii) 2 degrees centigrade for a highly optimistic scenario. The parameters constituting the scenario classes vary in both time and space. Again, even in the same climatic scenario, the increase in temperature will be greater over land, higher latitudes and winter compared to sea, lower latitude and summer respectively. The above may well account for differences in rate of increase of temperatures relative to 2000 reported by different workers for different areas and seasons such as an estimated increase of 2.5 °C by 2100 by Kattenberg et al., (1996) and by 2050 by Lal et al., (2001). In view of the above, IPCC (2007) has postulated that the rate of warming over each inhabitable continent is likely to be twice as large as that in the 20th century. Thus, for impact assessment studies the rate of increase may be taken as the most recently observed ones on a region-wise and season-wise basis.

Another interesting feature is that the rise in minimum temperature will be more than that of maximum (Karl et al., 1991) leading to a steady decrease in the daily temperature range over the years. This non-symmetry in increase of minimum and maximum temperatures has great agronomic significance which will be discussed in a subsequent section.

9.3.8   Evaporative Power of Air

Evaporative Power of Air, EPA is the capacity of air to desiccate moist surfaces given 100% opportunity. It is a principal factor determining water needs of crops. As mentioned in an earlier chapter Pan Evaporation, EP and Reference Crop Evapotranspiration ET are measures of EPA. In a study of evapotranspiration of Safflower at 2 locations in India, Venkataraman (1985) found that Cumulative Pan Evaporation, CPE could be related to Accumulated Heat Units above 0 °C, AHU > 0, at the rate of about 5 AHU > 0 to 1 mm of CPE. A close relationship between AHU and CPE has been noted by Sastry and Chakravarthy (1982). Thus, the 4th assessment report of IPCC (2007) has postulated that global warming will increase evaporative demand. The above assumption has been made by many workers also.

However reduction in computed values of ET over the last few decades in the world have been reported in many places like India (Chattopadhyaya and Hulme, 1997; Bandophadyaya et al., 2009) and China (Thomas, 2000; Gao et al., 2006; Li et al., 2008; Song et al., 2010, An and Li, 2005). The same trend in EP has been noted in India (Chattopadhyaya and Hulme, 1997), China (Liu et al., 2004; Chen et al., 2005, An Yuegai and Li Yuanhua, 2005), Australia (Roderik and Farquhar. 2004). New Zealand (Roderik and Farquhar, 2005), USA (Petersen et al. 1995) and in many parts of Russia (Golubev et al., 2001). The magnitude of trend in EP is larger than the IPCC estimate of radiative forcing of doubled CO_2 and opposite in direction (Roderik et al., 2009). The rates of decrease in ET varies with the seasons, being highest in spring and lowest in winter (Li et al., 2008).

The decrease in ET and EP despite an increase in temperature called the "Evaporation Paradox" (Golubev et al., 2001) for which an explanation in terms of the hydrological cycle has been offered by Brutsaert and Parlange (1998). Reduction in wind speed and sunshine hours(An and Li, 2005; Li et al., 2008) and lower Net Radiation and wind speed (Song et al., 2010) were seen as the main reason for decrease in ET and EP respectively. Influence of wind speed on ET or EP is very much limited compared to solar radiation, Decrease in EP can also arise on account of reduced bright hours of sunshine brought about by higher aerosol concentration (Roderik and Farquar, 2002). So the Evaporation Paradox seems to be due to reduction in solar radiation on account of cloudiness and/or higher aerosol concentration. The implications of the above findings on crop water needs will be discussed in a subsequent section.

9.4   Aerosols

An aerosol is a suspension of very fine solid particles or liquid droplets in a gaseous medium. Fog, smoke from forest and grassland fires, dust from volcanic eruptions and dust storms and salt from sea spray are examples of naturally occurring aerosols. Human activities like burning of fossil fuels and biomass, alteration of natural surface cover on earth and industrial activities also generate aerosols. Burning of fossil fuels and coal produce Sulphur which on oxidation produces sulfated aerosols. (Pruppacher and Kleet, 1978). Soot also referred to as Black Carbon (BC) can be of both natural and anthropogenic origin. Ten percent of the aerosols in the atmosphere is of anthropogenic origin. Most of the anthropogenic aerosols is concentrated in the Northern Hemisphere and in areas downwind of industrial areas, regions of shifting system of agriculture and in overgrazed grasslands. The concentration of natural aerosols in air has remained constant over long periods of time. The concentration of anthropogenic aerosols in air has been increasing since 1950.

Aerosols can cool the climate directly under clear sky conditions by direct scattering and reflecting away some of the incoming solar radiation and indirectly under cloudy conditions by increasing the reflectivity of clouds and thus offset global greenhouse warming to a considerable degree. Studying the direct radiative forcing of several natural and anthropogenic aerosols, Jacobson (2001) found chloride, natural sulfate, sea spray and black carbon to be the most important aerosols. Others had also found sulfates and black carbon, both of anthropogenic origin, to play a vital role in regulating climate change

(Santer et al., 1995). Sulfate aerosols act as cloud condensation nuclei leading to increased adherence of cloud droplets and a brighter cloud (Twoney, 1974) and increase the planetary albedo. In the Northern Hemisphere, sulfate aerosols exert a cooling effect that is comparable in magnitude but opposite in sign to the warming of greenhouse gases (Charlson et al., 1992). Taylor and Penner (1994) incorporated the effects of Sulfate aerosols in coupled, ocean-atmosphere general circulation model to study the patterns of response in surface temperature. Model runs with increasing CO_2 or sulfate aerosols as the only dominant influence on climate, generated strong multi-decadal warming and cooling trends respectively which were not in tune with recorded observations. Combining both CO_2 and anthropogenic sulfate aerosols in the model resulted in generation of results in closer agreement with observed past temperatures (Santer et al., 1995) and spatially coherent warming and cooling data such as lower temperatures of the northern hemisphere. While aerosol cooling in winter will offset warming due to CO_2, it will, in Summer, weaken the Asian Summer monsoon circulation (Mitchell and Johns, 1997) by weakening the land-sea temperature contrast. The role of sulfate aerosols in multi-decadal fluctuations in climate is larger than indicated by climate models. Black carbon aerosols, by reducing (i) albedos of cloud particles and other aerosols and (ii) cloud cover, cause warming. They can directly absorb solar radiation, heat the air and cool the surface. Largely due to growth and coagulation of aerosol particles, Black Carbon can get incorporated within the aerosols and as an internal mixture, exert a higher positive forcing greater than that of Methane (Jacobson, 2001) and contribute significantly to global warming. However when soot is deposited on a surface, it lowers its albedo leading to heating of the surface.

Balancing the effects of Green House gases by aerosols is not possible. The distribution and concentration of aerosols vary in time and space. So their mitigating influence will be sporadic, variable and local. Further aerosols have a life of one to two weeks compared to 100 years of Carbon Dioxide. This means that aerosols need to be continuously generated to mitigate influence of GHGs. The aerosols are also air pollutants. Sulfate aerosols cause acid rain, lung irritation and haze. The brown colour of the Asian Brown cloud is predominantly due to Black carbon (Srinivasan and Gadgil, 2002). Checking emission of anthropogenic aerosols is necessary for both developing and developed countries to prevent encroachment of warm cloudy weather into present areas and seasons of bright cool weather (Venkataraman, 2003). Increasing aerosol burden from 1950s to 1980s has led to Solar Dimming i.e., lesser solar radiation reaching the surface while decreasing aerosol burden from 1980 to 2000 has led to Solar Brightening i.e., more solar radiation reaching the surface (Wild et al., 2007). The implication is that reduction in emission of sulfate aerosols and soot, so necessary in the coming years, would call for redoubled efforts to reduce emission of GHGs.

9.5   Climate Change Controversies

In the field of climate change there are two opposing schools of view advocating cognizance and ignoring of climate change. Some of the arguments advanced by both schools of view are flawed while some others merit consideration. The above dichotomy arises from the fact that a change in a given weather element does not occur in isolation but is reflected in changes of other parameters and the effect of combination of changes on any physical or biotic entity/process is quite different from what would result for a given parametric change. The present spatial variations of climate on earth are due to coverage of the planet with different types of surfaces, each with its own radiative temperature even under the same input of energy. Thus, any large scale change in the type of surface, like deforestation, will bring about a local climate change. Again, pollution of soil and water resources brings about a dwindling down of aquatic fauna and terrestrial flora. For any weather parameter, there are decades of epochs of increase and decrease, which will affect biotic processes in a manner similar to that of climate change. However, climate change will be a super imposition on natural weather vagaries and effects of anthropogenic follies and will have a significant and visible impact on crops and affect large masses of people. Keeping the above in view we may examine a few of the arguments advanced for neglect and recognition of climate change.

Valid arguments advanced for neglect of climate change are the following namely:

(a)   Climate change predictions are based on simulation models. Since we are unable to predict either the temporal or spatial vagaries of weather in a season, not much store can be set on prediction of weather 50 or 100 years from now.

(b) The magnitude in changes of weather elements envisaged even in the pessimistic climate scenario do occur even now at a place either inter or intra seasonally and

(c) Technology developed to cope with current level of weather vagaries can be used to cope with future climate change scenario.

The criticism relating to use of models as at "a" above is valid. Thus, for assessing impact of climate change it is better to take, on a region-wise and season-wise basis, recently observed rate of change of parameters for the business as usual climatic scenario, twice the current rate of change for the pessimistic climatic scenario and 50% of the current rate for the most optimistic scenario.

In the temporal march of weather elements, to which "b" above refers, periods of aberrant weather are limited to a short period and the effects of one period of weather vagary can be offset by another period of opposing trend as in the case of rainfall and temperature. However, when there is an overall increase in temperature the crop life-duration will get shortened, similarly increases or decreases in rainfall will lengthen and curtail crop life-durations respectively. In other words, influence of climate change will get superimposed on that of aberrant weather and the two effects may be additive or opposing. The above argument also applies for epochs of increasing or decreasing trends of a weather parameter. So far as carbon dioxide is concerned, occurrence of higher levels in the past cannot be pleaded and the level will keep on building with time. The argument that increased CO_2 will have a fertilising effect on crops and boost up yields ignores the fact that such an increase will be accompanied by an increase in temperature and cloudiness which will (i) tend to offset the beneficial effects of enhanced CO_2 and (ii) may lead to encroachment of warm, cloudy weather into current periods and regions of cool, clear weather.

Regarding use of current technologies as mentioned at "C" above, new technologies. Both agronomic and genetic, will be required, as detailed under the discussions on mitigation of climate change, to combat effects of climate change.

Citation of events and unusual changes as evidence of climate change by its protagonists are not without flaws as detailed below.

(i)   *Early, flowering of tree species*: One is not too sure if this is due to occurrence of epochs of increasing temperatures. It would be better to study flowering behaviour of plants with long gestation periods in flowering like Bamboo in Assam and Kurunji in southern Tamil Nadu, India.

(ii) *Absence of pre-monsoon and post-monsoon hail storms and reduction in thunderstorms*: These weather vagaries could also be ascribed to the diminution of natural vegetation in rural areas to make way for agricultural use resulting in barren ground cover in summer

(iii) *Warming of metropolitan areas*: This is mostly due to construction of concrete cement buildings.

(iv) *Loss of some aquatic, terrestrial and avian species*: It is now well known that surface runoff from rains often carries residues of agrochemicals along with it. Industrial effluents containing harmful chemicals are allowed to be discharged in water reservoirs subject to the chemicals being below specified concentration levels. In the natural food chain the concentration of the discharged chemicals keeps on increasing and reaches a value which is lethal to those higher in the food chain. For example, in the food chain consisting of phyto-plankton, zoo-plankton, small fish, large fish, snakes and vultures, the increasing concentration becomes lethal for vultures leading to its disappearance over such water bodies. Extinction of species is, therefore, not due to climate change but is a result of human intervention.

However, notwithstanding the above discussions, it is better to be prepared for climate change for the simple reason that injection of large volumes of CHGs is bound to interfere in the energy balance of the earth-atmosphere system and warm the earth's surface. There is enough evidence by way of melting of polar ice caps, retreat of glaciers etc., to confirm that global warming and climate change are taking place. The view that the terrestrial biosphere can act as an overall sink for CHGs till 2050 (Cox et al., 2000) appears misplaced and the time to act for mitigation, if not elimination, of the effects of the ongoing specter of climate change on crop production is now.

9.6 Effects of Climate Change on Crops

9.6.1 Elevated Carbon Dioxide

Elevated levels of CO_2 increase leaf area development (Allen, 1990), which per se leads to increased water requirements on a unit land area basis. Higher CO_2 levels also lead to a decrease of stomatal aperture which reduces transpiration (Goudrian and Unsworth, 1990).

The reduction in stomatal conductance for a doubling of CO_2 averages about 34% (Cure and Acock, 1986). The reduction in transpiration for a doubling of CO_2 ranges from 23% (Cure and Acock, 1986) to 40% (Morison, 1987). Thus, elevated CO_2 will help crops to cope with drought conditions better (Samarkoon and Gifford, 1995).

The reduction in transpiration under elevated CO_2 will increase leaf surface temperature and hence crop canopy temperature. Work relating canopy temperature based indices to yield by several workers shows that grain yields of many crops will improve with an increase in canopy temperature above the ambient air temperatures.

The lowered stomatal conductance due to enhanced CO_2 does not interfere with gas exchanges between leaf and air and photosynthesis is thus not affected (Drake et al., 1997). Therefore, elevated CO_2 will increase photosynthesis (Allen, 1990).The indications are that, on an average. A doubling of CO_2 will increase yields of many crops by 30% (Kimball, 1983). Current levels of CO_2 is a limiting factor in photosynthesis of C3 plants (Pearcy and Bjorkman, 1983). Elevated CO_2 will increase dark respiration (Drake et al., 1997) but the ratio of dark respiration to photosynthesis is constant under CO_2 enrichment (Casella and Soussana, 1997). Thus, elevated CO_2 will reduce photo-respiration, which will benefit C3 plants more than C4 plants (Johnson et al., 1993). Most of the weeds are C3 plants. Thus, elevated CO_2 will lead to a smothering of C4 Crops by C3 weeds. Higher CO_2 will benefit trees and shrubs more than grasses leading to woodlands expansion and shrubs encroachment (Krishan and Saha, 2008). Elevated CO_2 increases photosynthetic rates and plant growth of natural ecosystems (Bazzaz, 1990) and of fruit trees like Sour Orange (Idso and Kimbal, 2001), Citrus (Keutgen and Chen, 2001) and Apple (Pan et al., 1998). Elevated CO_2 is more beneficial for plants that fix Nitrogen (Cure et al., 1988; Ainsworth et al., 2002).

For a doubling of CO_2 increase in yield varies amongst C3 and C4 crops, being (a) 9% for maize(Patel et al., 2008) (b) 15% and 28% respectively for rice and wheat (Lal et al., 1998) and 50% for soybean (Lal et al., 1999). 50% increase in CO_2 increases rice yield by 7% (Hundal and Kaur, 1996) but does not affect productivity, photosynthesis and yield of maize (Leakey et al., 2006).

Under elevated CO_2 indications are (Cure and Acock, 1986) that after acclimatisation, the initial high rates of photosynthesis will begin to decline to a steady value. Kimball and Idso (2005) report that sour oranges exposed to CO_2 concentrations of 300 ppm over the normal produced 3 times more biomass than the control in the first 2 years after which the rate of increase in biomass began to decline to a steady value of 1.7 times the control from the 9th year.

Since elevated CO_2 leads to increase in biomass production with an accompanied decrease in water needs, water use efficiency in biomass production increases with increase in CO_2 levels. C3 plants may far better than C4 plants under soil moisture stress conditions on account of their improved water use efficiency (Owensby et.al. 1993).

Elevated CO_2 (i) reduces nutrient uptake and nutrient utilisation efficiency (Roger et al., 1994; 1999). Due to inadequate nutrient absorption, plants are unable to attain the potential increased shoot growth under elevated CO_2 (Bassirirad et al., 1997). Higher CO_2 increases (ii) root to shoot ratio and contributes to build up of organic matter in soil (Mauney et al., 1992) and (iii) tillering of crops. Doubling of CO_2 increases yield of rice by 32% due to increased tillers and panicles per plant (Baker et al., 1990). Increase in CO_2 levels lead to an increase in carbon to nitrogen ratios in plant tissues and a decrease in the Nitrogen and protein content of cereal grains (Fangmeier et al., 1999).

There are some indications that elevated CO_2 may have some effects on pests and diseases of crops. Because of heavier foliage development, higher CO_2 is held to favour pathogens. But the reduction in transpiration and the resultant increase in leaf temperatures will countenance the favourable effect of a denser canopy. High CO_2 can improve resistance of hosts to pathogens (Coakley et al., 1999). Longer and increased availability of crop residues on account of their lower rate of decomposition under elevated CO_2 can help in easier over-wintering of pathogens, favouring earlier and rapid development of diseases.

Elevated CO_2 can indirectly influence incidence of insects through its effect on the host plant. Increase in levels of simple sugars in soybean leaves under elevated CO_2 is thought to be a contributory factor in increased damage by many insects (Hamilton et al., 2005). The lowered nitrogen content of leaves under elevated CO_2 can lead to voracious consumption of leaves by insects to meet their metabolic requirements of nitrogen (Hunter, 2001) and may retard development of insects and increase the length of the damaging life stages of insects (Coviella and Trumble, 1999).

The main benefits that can accrue from elevated levels of Carbon Dioxide are increased yields and reduced water requirements of crops. Lowering of stomatal conductance will increase surface temperature of leaves. As mentioned in the following section, increase in CO_2 levels are accompanied by increase in air temperatures which reduce field-life duration. This will reduce yield of crops and offset the increase in yields from CO_2 enrichment. Thus, the main tangible benefit due to increased levels of CO_2 is the reduction in transpirational needs of crops.

9.6.2 Higher Temperatures

The conservative nature of yield per day of field-life duration for a given cultivar in a given season has been mentioned earlier. Field-life duration and durations of various phenological phases of a crop cultivar is determined by its accumulation of degree days over a specified base temperature. Thus, any increase in temperature will reduce the field-life duration of a cultivar and reduce its potential yield. The base temperature for accumulation of day degrees varies amongst crops. As mentioned earlier, for given cultivar and a given temperature increase, the extent of reduction in field-life duration will vary with the level of ambient air temperatures. Thus, a given rise in temperature will affect reduction in life-duration of (i) different crops differently in the same location (ii) of the same crop differently in different locations in the same season and (iii) of same crop in different seasons in the same location. It is this feature that necessitates assessment of impact of climate change on crop prospects to be carried out on a crop-wise, location-wise and season-wise basis. The last aspect can be taken care of by the use of mean crop seasonal ambient temperatures of 15, 20, 25 and to 30 °C (Venkataraman, 2004).

As higher temperatures are due to elevated CO_2 levels, Dynamic Crop Weather simulation models have been used to study the combined effects of elevated CO_2 and the associated temperature increase on the yields of many crops at many diverse environments (Crisanto and Leandro, 1994; Hundal and Kaur, 1996; Lal et al., 1998; 1999; Aggarwal and Mall, 2002; Attri and Rathore, 2003). Such studies show that the increase in temperature needed to offset a doubling of CO_2 concentration will vary with location, season and crop. However, when the temperature increases to 3 degrees centigrade above normal, the beneficial effects of doubled CO_2 gets canceled out for any crop in any season and any location. As mentioned earlier, doubling of CO_2 will lead to an increase of temperature from 3 to 5 degrees centigrade. Thus, in real-time increase in elevated CO_2 will lead to decline in crop yields (Sinha and Swaminathan, 1991).

As mentioned earlier, in global warming, minimum temperatures will increase more than the maximum. For most plants photosynthetic rates remain unaffected over a wide temperature range. For example photosynthesis in rice is little affected over temperature range of 20 to 37 °C (Penning de Vries et al., 1989). Increase in minimum temperature will lead to increased respiratory depletion of photosynthates. The maintenance respiration for rice is 7% of photosynthesis (Mall and Aggarwal, 2002) and about 10% of photosynthesis for other crops (Bishnoi, 1986). The Q10 for maintenance respiration is 2.0 with a reference temperature of 25 °C (Penning de Vries et al., 1989). Thus, increased temperature will reduce biomass production in crops. Increased temperature will also increase the photo-respiration of C3 crops (Long, 1991). Thus, the effects of increased temperature on yield reduction will be more for C3 than C4 crops.

Since snow reflects 95% of the radiation, melting of snow occurs with a rise in ambient air temperature. Thus, global warming will lead to melting and reduction in extent of snow cover. Even now glaciers are reported to be melting and shrinking. Thus, in global warming in irrigation systems dependent on water from snow-melt, there will be an increase in quantum of water in the irrigation system initially. The associated increase in thickness of fresh water layer in the oceans in deltaic areas will lead to an increase rainfall in coastal areas. This will give a false sense of increased availability and water-security. However, with progress of time coastal rainfall and water availability will begin to decrease at an alarming rate leading to reduction of crop production at an increasing rate in such systems.

Increased temperatures will lead to a decrease in response of crops to added fertilisers. Thus reinforcing the effects of elevated CO_2 on reduction in uptake and utilisation of nutrients (Rogers et al., 1994). Higher air temperatures will lead to higher soil temperature, which will increase mineralisation and gaseous loss of Nitrogen. Thus, global warming will necessitate increased applications of nitrogenous fertilisers (Aggarwal and Mall, 2002; Aggarwal. 2003). This in turn will lead to greater infusion into the atmosphere of Nitrous Oxide.

Lower temperatures at present inhibit the northern and southern spread of weeds in the northern and southern hemisphere respectively. Thus, global warming will lead to an increase in weed infected areas. The beneficial effects of increased biomass accumulation in C3 plants will be nullified by the increased temperatures associated with elevated CO_2 and there will be no predominance in the spread of C3 weeds.

Currently temperatures in India are (a) one degree centigrade above optimal for rice and wheat (Lal et al., 1998) and (b) 3 degrees above optimal for Soybean (Lal et al., 1999). Thus, global warming will adversely affect crop prospects much more in low latitude tropics than in temperate regions. Initially the increase in temperatures will lead to an increase in frost-free period and earlier maturity and harvest of crops in temperate zones and benefit agricultural and forage crops. There will be reduced and/or less severe incidence of frosts in all crop-frost prone areas. With passage of time crop prospects in temperate zones will also begin to worsen under global warming.

Higher temperatures will lead to a faster development of pests and diseases. However, as host plants will also be developing faster under increased temperatures, the duration of availability of susceptible crop stages to pests and diseases may not change significantly. Despite this, rise in temperature will have significant effects on the incidence, severity and spread of pests and diseases as detailed below.

Global warming will render over-wintering of pests and diseases easier. Higher populations of pests and pathogens will survive for a greater initial infection of crop plants. Higher temperatures will lead to earlier appearance and more number of generations of pests and diseases in a season. This will result in northward and southward spread of pests and diseases in the Northern and Southern Hemisphere respectively. Northward spread of (i) Cottony Cushion Scale in Europe (ii) Cottony Camellia Scale in U.K. and (iii) Oak Moth in Central and Southern Europe due to higher temperatures (FAO, 2008) and poleward extension of several crop pests in Japan in the period 1965-2000 (Kiratani, 2007) are cases in point. Migratory pests like Corn Earworm will arrive earlier at a location or the area in which they are able to over-winter will increase.

Warmer temperatures favour and inhibit respectively thermophilic and thermophobic fungi. Rise in temperatures will respectively increase and decrease Black Rust (Puccinia graminis tritici) and Yellow Rust (Puccinia striiformis) of wheat. Wheat and Oats become more susceptible to themophilic rusts with increase in temperatures (Coakley et al. 1999). However, rice becomes susceptible to the Blast disease (Pyricularia oryzae) under low temperatures of 25 degrees centigrade or below (Sadasivan et al., 1963). Black Rust of wheat will appear earlier in the wheat belt of plains of North India. Incidence of Late Blight (Phytophthora infestans) is earlier and/or more severe under higher temperatures (Wallin and Waggoner, 1990). In Finland and Northeastern USA, one degree centigrade rise in temperature results in earlier appearance of late blight by 4 to 7 days (Hijmans, 2000) with an extended susceptibility period by 10 to 20 days (Kaukoranta, 1996). Higher temperatures will lead to lesser leaf-wetness duration due to higher nocturnal crop-leaf temperatures and reduce risk of incidence of many diseases.

Lower mortality of insects due to warmer temperatures will lead to an increase in the population of insects (Harrington et al., 2001). In nature diversity of insect species per unit area decreases with an increase in altitude and latitude (Andrew and Hughes, 2005). Thus, increase in temperatures will lead to more pests attacking more crops in the temperate climatic zones (Bale et al., 2002).

Some vectors and natural enemies of pests and diseases do better under higher temperatures. Some Aphid vectors will arrive earlier or become more active (Harrington et al., 2007). Corn Flea beetle

(Chaetocnema pulicaria) is the vector for the Bacterial Blight (Erwinia stewartii) and its survival is greater at higher temperatures. Extent of survival of the vector determines severity Stewart's wilt infections in the following year, which can be forecast from winter temperature regimes (Castor et al., 1975). Higher temperatures can cause an explosive rise in population of Ladybirds (Adalia Sp), which are predatory on Aphids (Majerus and Kearns, 1989). At higher temperatures Aphids become less sensitive to the pheromones they release when under attack by enemies and thus become liable to greater predation and parasitization (Awmack et al., 1997). Higher temperatures often offset the favourable effects of CO_2 on pests and diseases such as the Cereal Aphid (Newman, 2004). Higher temperatures can change the gender ratio of some pests like thrips (Lewis, 1997) and reduce effectiveness of some classes of pesticides like Pyrethroids (Musser and Shelton, 2005).

9.6.3   Reduced Solar Radiation and Increased Cloudiness

The relationships of solar radiation and crop yields and effects and ways of coping with reduced solar radiation and increased cloudiness envisaged under climate change have been dealt with in some detail in Chapter 4 on "The Concept and Relevance of Radiation Balance in Crop Culture".

9.6.4   Lower Evaporative Power of Air

The combined effects of elevated CO_2 and increased temperatures will lead to a decrease in unit area yields of clear season irrigated crops (Sinha and Swaminathan, 1991). Thus, the need for increasing both net and gross irrigated area becomes urgent. Increase in net irrigated acreage by construction of dams across rivers is time consuming with a limited potential of 40% in India as a short term measure (Garg and Hassan, 2007). Increase in rainfall, some of which at the expense of solid precipitation is envisaged under global warming. However, water availability in irrigation systems dependent on snow/ice melt will decrease in the long run. Again, interlinking of national rivers to even out floods and droughts will not solve the irrigation problems of elevated regions like plateaux. So avoidance of waste of irrigation water is necessary to increase gross irrigated acreage by relay cropping.

By middle of this century plant-usable CO_2 will increase by 50%. (Sinha, 1993). This should lead to a reduction in transpiration demand of crops by 15%. Evaporative Power of Air can be expected to decrease by 10% by 2050. Thus, water needs of crops will be lesser by 25% and irrigated acreage can be increased by 33% with existing irrigation potential. With a 20% saving in surface irrigation, which is easily possible, the irrigated acreage can be increased by 70%. This will reduce the need for urgency and quantum of creation of additional irrigation potential to meet crop production needs.

9.7   Assessment of Crop Prospects in Future Climate

9.7.1   Climatic Scenario

There is no uniformity of opinion regarding the likely quantum and/or rate of change of temperature, CO_2 and solar radiation. Thus, for evaluating the impact of climate change on yield prospects of irrigated crops, it is first of all necessary to adopt three types of climatic scenario, namely: the optimistic, pessimistic and realistic ones and assign to each scenario class, specified but different (a) increases in temperature and CO_2 levels and (b) decrease in solar radiation. For example one can specify (a) CO_2 level of 460 ppm, 1 °C rise in mean temperature and 5% reduction in solar radiation for the optimistic scenario (b) CO_2 level of 550 ppm, 2 °C rise in mean temperature and a reduction of 10% of solar radiation for the realistic scenario and (c) 660 ppm of CO_2, 3 °C rise in temperature and 15% reduction in solar radiation for the pessimistic scenario.

9.7.2   Influence of Crops and Seasons

As mentioned earlier, crops vary in (i) the extent of reduction in field-life duration due to variations in (a) base temperature for growth and (b) level of ambient mean air temperature in their growing seasons and areas. Therefore, impact assessments of climate change on crop yields have to be carried out on a crop-wise, region-wise and season-wise basis.

9.7.3   Holistic Approach

There are differences amongst crops in (i) their efficiency in using increased CO_2 (ii) ability in coping with reduced solar radiation and (iii) their Q_{10} values for maintenance respiration. Thus, in studies on impact of climate change on crop prospects, it is necessary to adopt a holistic approach as detailed below. For any given crop season and region and climate scenario estimates for (i) increase in yield due to increase in level of CO_2 (ii) reduction in yield due to (a) reduction in radiation (b) reduction in field-life due to increase in temperature and (iii) decrease in yield due increase in maintenance respiration on account of increased temperature has to be worked out. The changes have then to be added to get the net change in yield (Venkataraman, 2004). There is a school of thought that the effects of environmental changes will be multiplicative instead of additive.

9.7.4   Role of Dynamic Crop-Weather Simulation Models

Under the auspices of U.S. Aid for International Development (USAID), Crop Environment and Resource Synthesis (CERES) models, sharing same data inputs and outputs and embedded in a software package called Decision Support System for Agrotechnology Transfer (DSSAT) had been developed by the International Benchmark Sites Network for Agrotechnology (IBSNAT) for a number of crops. DSSAT allows one to (i) run crop models with manipulated combinations of crop, soil and weather data (ii) simulate, on a daily basis, the effects of variations in weather and crop varieties on phenological development, gross and net photosynthesis and partitioning of net biomass accumulation into economic yield and (iii) analyse results emanating thereof. Such models have been developed for many crop species and can be classed as Dynamic Crop Weather Simulation Models (DCWSMs). Sub-models (i) on soil moisture balance for use with rainfed and irrigated crops and (ii) to simulate the effects of nitrogenous fertilisers on crop growth and yield are included in many DCWSMs.

9.7.5   Potential and Relative Yields

The effects of pests and diseases, hazardous weather phenomena and adequacy of mineral nutrition on crop yields are not considered in DCWSMs. So studies on impact of climate change on crop prospects by DCWSMs must perforce be limited to irrigate adequately nourished and well protected crop cultivars growing in an environment free of hazardous weather phenomena so that only temperature, solar radiation and CO_2 levels are the operative factors determining their yield level. Thus, yields obtained by use of DCWSMs will refer to Potential yields obtainable under a given weather situation in the crop season. However, potential yield of an irrigated crop in the same location and season will vary with crop varieties. Therefore, field-validation of DCWSMs require crop coefficients for determining phenological development, photosynthesis, respiration, net biomass production and partitioning of biomass to economic yield. Thus, outputs of DCWSMS are cultivar-specific. However, relative changes from potential crop yields can be expected to hold across varieties of a given crop at a given location in a given season.

9.7.5.1 Irrigated Crops

Assessments of impact of climate change on yield prospects of irrigated crops on a crop-wise, region-wise and season-wise basis under optimistic, pessimistic and realistic climate scenarios by DCWSMs and the data inputs needed to operate the models make the task an onerous one. However, such studies by use of DCWSMs can indicate, for a given, changed climatic scenario, the relative change in potential productions of a crop cultivar, appropriate to the location and season. For real-time use, such relative assessments would suffice.

9.7.5.2 Dryland Crops

For assessment of yield prospects of dryland crops, one needs to know not only the quantum of rainfall but also its temporal distribution, both of which cannot be deduced from available General Circulation Models (GCMs). However, since an increase in rainfall due to global warming has been indicated and since, as mentioned earlier, dryland crops on account reduced transpiration due to elevated Carbon Dioxide can withstand drought better, one can only say that dryland crop prospects may not drastically worsen in future.

9.7.5.3 Crop Field-Life Duration

Since base temperatures for almost all crops are known, it is simple to calculate the percentage reduction in field-life duration of a crop cultivar for a given ambient seasonal mean temperature and hence the percentage reduction in unit area yields, The percentage reductions in yields thus deduced for a given change in temperature will hold for different cultivars of a given crop at a given location and season.

9.8   Need for and Role of Controlled Environmental Facilities

Decrease in solar radiation, as envisaged in climate change scenarios, occur in real-time crop culture on an inter-daily or inter-regional or inter-seasonal basis. However, such a situation does not obtain for carbon dioxide. So for studying effects of increased CO_2 levels on photosynthesis, growing of plants in controlled environment with facilities for maintenance of specified CO_2 levels has to be resorted to. Besides the well known Phytotrons, such facilities described by Uprey et al., (2006) include Open Top Chambers, Sunlit Controlled Environmental Chambers, Screen-aided CO_2 Control, Portable Field Chambers, Free Air Temperature Enrichment (FATE) and Free Air Carbon Dioxide Enrichment (FACE) facilities and Soil Plant Atmosphere Research (SPAR) Units. This allows recourse to be taken on field measurements of canopy photosynthesis at different levels of temperature and CO_2.

9.8.1   Influence of Solar Radiation

Assessment of decrease in crop yields due to the envisaged decrease in solar radiation is relatively simpler. The SPAR units provide (i) 95% transmissibility to PAR (ii) control of soil and aerial environmental factors over a wide range and (iii) continuous measurement of gross and net canopy photosynthesis of plants grown at optimum, maximum and minimum temperatures and ambient CO_2. For natural values of solar radiation, canopy gross photosynthesis can be calculated (Reddy et al., 1995) to establish a relationship between the above two parameters. Assuming a limiting value of Solar Radiation of 25 $MJ/m^2/day$ for potential photosynthesis, such relations can be used to determine the extent of reduction in potential photosynthesis on account of actual amount of solar radiation received and develop factors for reduction in potential yields in terms of reduction in radiation vis-a-vis the value for 25 $MJ/m^2/day$.

9.8.2   Influence of Elevated CO_2

All climate change scenarios envisage increase in CO_2, albeit to various levels. Thus, in controlled environment studies on crops, it is necessary to artificially increase CO_2 over ambient levels and maintain such levels with accuracy during the course of field trials. The Free Air Carbon Dioxide Enrichment (FACE) facilities and Soil Plant Atmosphere Research (SPAR) Units provide the means to study CO_2 enrichment on crops. Of the two, maintenance of sub-ambient and super-ambient levels of CO_2 is held to be superior in the SPAR system.

There is evidence (Reddy et al., 1998) that in case of Cotton, CO_2 enrichment does not affect the optimum temperatures for vegetative and reproductive growth or stimulation of growth across temperatures as revealed by the uniformity of ratio of dry weight at 700 ppm CO_2 to that at 350 ppm CO_2. The above features simplify studies on CO_2 enrichment on crop photosynthesis.

In a manner similar to that of solar radiation, relationships can be established between canopy photosynthesis and CO_2 levels. In DCWSMs the effect of CO_2 level on net photosynthesis is simulated by multiplying the net rate by a factor based on studies of ratios of photosynthesis at a given level of CO_2 to the normal level of 330 ppm (Mall and Aggarwal, 2002). Studies of Peart et al., (1988) on effects of CO_2 levels on many C3 and C4 crops show a linear increase in ratio from 1.0 at 330 ppm to 1.25 and 1.43 at 660 ppm and 990 ppm respectively. The photosynthesis ratios of 550 ppm CO_2 to 330 ppm CO_2 are seen to be 1.21, 1.17 and 1.06 respectively for Soybean, Wheat and Rice (Allen et al., 1987; Cure and Acock, 1986 and Kimball, 1983). For C3 crop plants, the ratio of photosynthesis at 900 ppm CO_2 to 360 ppm CO_2 was seen to be 1.37 (Kimball, 1983; Reddy and Hodges, 2000). Idso and Idso (2000) report (i) a near uniform increase of 70% for a doubling of CO_2 from the normal level of 330 ppm and (ii) on an average, a 300 ppm increase in atmospheric CO_2 enrichment leads to yield increases of 15% for CAM crops, 49% for C3 cereals, 20% for C4 cereals, 24% for fruits and melons, 44% for legumes, 48% for roots and tubers and 37% for vegetables.

9.9   Mitigation of Agricultural Effects on Climate Change

Agriculture is a significant (i) source of CO_2, Nitrous Oxide. N_2O and Methane, CH_4 and (ii) sink for CO_2. So agricultural mitigation measures vis-a-vis climate change would consist of (a) measures for (i) lowering emissions of N_2O. CH_4 and CO_2 (ii) replacement of one emission by another one with less Global Warming Potential (GWP). Example: N_2O by Methane and Methane by CO_2 (iii) creation of additional sinks for absorption of CO_2 (iv) sequestering of carbon in soils (b) genetical manipulations to evolve cultivars to (i) cope with higher temperatures and lower radiation and (ii) take advantage of increased CO_2 levels (c) agronomic field measures to adjust to changed thermal and radiation regimes and (d) increasing sink capacity for CO_2 through optimal use of irrigation water to increase occupancy of land by crops in time and space.

9.9.1   Reducing Methane and Nitrous Oxide Emissions from Rice Culture

Rice cultivation under flooded conditions is a major source of atmospheric Methane. In view of the increasing needs of rice as a staple food, reduction in rice acreage to reduce methane emissions is not an option. Considering that a major fraction of water used in rice culture is due to non-crop usage, attempts are afoot to bring about significant reduction in use of water for raising a rice crop. However, any water that may be saved is likely to be used to locally increase rice acreage and thus lead to increased emission of Methane into the atmosphere. Methane flux from rice is affected by water management, incorporation of organic matter, organic carbon content of soils, soil pH and climate (Yan et al., 2005). Thus, in situ reduction of methane from flooded rice fields is a must and has rightly been attempted. The mitigation options have related to (a) manipulation of (i) irrigation (ii) Nitrogen fertilisation (iii) organic matter management and (iv) tillage (b) use of soil amendments and (c) selection of varieties. The main indications that emerge from such efforts are briefly mentioned below.

9.9.1.1 Altering Water Management

Draining of puddled rice once or several times during the growing season has been resorted to for reducing methane emissions through prevention of development of soil reductive conditions. Of all forms and permutations of draining of water, aeration provided by multiple drainage emerges as the most effective in reducing methane emissions. The extent of reported reductions range from 88% (Sass et al., 1992) to 40% (Tyagi et al., 2010) with an average of 50% in Asian countries (Yan et al., 2005). Reduction in methane emission persists for quite some time after re-flooding of the field (Yagi et al., 1996).

However, the above reductions in methane emission from rice field is to some extent offset by an increase in emission of nitrous oxide (Akiyama et al., 2005) due to aeration occurring in the time between draining and re-flooding of the rice field. Nitrous oxide emission was influenced by drainage day rather than number of times water is drained. Shorter drainage time was required for reducing methane emissions (Towprayoon et al., 2003). There seems to be a trade-off between methane and nitrous oxide emissions in all treatments, with treatments reducing methane emissions increasing nitrous oxide emissions. About 15 to 20% of the benefit gained by reduction in Methane emission is offset by the increase in Nitrous oxide (Yan et al., 2005). Since, molecule for molecule, Nitrous Oxide is ninety times more effective than Methane in trapping heat, the contribution of Methane and Nitrous oxide in various drainage systems in terms of Global Warming Potential (GWP) needs an examination. It has been reported (Jiang et al., 2006) that from a flooded rice field, methane contributes 93% of GWP. Thus, even with increased Nitrous oxide emissions, multiple drainage is held to be an effective option for mitigating net GWP from rice fields.

It would appear that with use of same quantum of water as that required for flooded rice, multiple drainage invariably results in about 10% yield reduction of rice vis-a-vis the flooded crop. (Towprayoon, 2003). This is not surprising as maintenance of soil water content at slightly above field capacity leads to a 20-25% reduction in yield compared to flooded rice (De Datta, 1981). Irrigating to maintain soil water status at 0.15 bar matric potential had been reported (Khosla et al., 2011) to lead to 60% reduction in emission of methane compared to flooded rice with no decrease in yield. No mention about the quantum of water used to maintain the 0.15 bar matric potential has been mentioned. Saas et al., (1992) report that under multiple aeration, to maintain yields as at flooded status required 2.7 times more water compared to the flood water treatment.

9.9.1.2 Fertiliser Applications

Pure mineral fertilisers and pure non fermented organic fertilisers produce respectively the lowest and highest emissions of methane. Fermented, organic fertilisers like sludge from biogas generators led to substantially lower methane emissions due to depletion of potential methane precursors during fermentation (Wasmann et al., 1993). Composting or incorporation of organic matter in the off season drained period (Yagi et al., 1997)) is recommended to reduce methane emissions. Sulfate-containing soil amendments are also advocated as a mitigation option to reduce methane emissions from rice fields. The amount of reduction in methane depends on quantum of sulfate applied and is location-specific (Denier Van Der Gon et al., 2001). Comparative studies show that ammonium sulphate, which favours activity of other microbial groups over that of the methanongens., gave 50% more potential reduction in methane emissions than urea (Schutz et al., 1989), while calcium sulfate anhydride has a 10% more potential for reduction of Methane than ammonium sulfate (Metra-Corton et al., 2000). In both organic and inorganic mineral nutrition of flooded rice, there seems to be a trade-off between emissions of methane and nitrous oxide. For example application of Urea while tending to reduce methane emissions significantly, increases nitrous oxide emissions (Zou et al., 2005). Similarly addition of Biochar increased methane emissions but reduced that of nitrous oxide (Zhang et al., 2010). However, quantification in terms of GWP in such trade-offs are not precisely available.

In view of the above, amendments that reduce both methane and nitrous oxide emissions become important. Use of Calcium Carbide (Bronson and Mosier, 1992) and Dicyanamide (Xu et al., 2002) to reduce emission of both CH_4 and N_2O in rice is noteworthy.

9.9.1.3 Choice of Cultivars

Methane emission in rice is influenced by growth stages of the crop and generally two to three peaks of emissions are observed in the rice growing season. Under the same growing conditions of soil and climate, rice cultivars show wide variations in unit area emissions of methane per day. The range is from 35 to 250 mg/sq. m./day. The rate decreases with increasing duration of crop and the highest increase occurs with high yielding/hybrid varieties (Reddy et al., 2005). Cultivars with small root systems, high root oxidative activity, high harvest indices and more tillers are likely to produce less methane than other cultivars (Wang and Adachi, 2000). Use of cultivars with low exudation rates (Aulakh et al., 2001) and minimum number of tillers but with maximum percentage of productive tillers (Setyanto et al., 2004) can reduce methane emission from rice. Biomass content has a significant influence on methane emission as the aerenchyma cells of rice plant which act as conduits for release of methane to the atmosphere are an integral part of rice (Reddy et al., 2005).

Studies conducted in China (Wang et al., 2000) and in Korea (Shin and Yun, 2000) confirm that seasonal fluxes of methane can vary by a factor of two. Thus, choice of rice cultivars offers a practical methane-mitigation option as no change in existing practices are called for. Change of cultivars, however, results in less significant reductions of methane than other mitigation options (Lu et al., 2000). Again, it seems from the above that the greater the seasonal biomass production the greater will be the seasonal integrated flux of methane, thus, it is necessary to ensure that use of the low-emitting cultivars does not result in reduced yields. In this, the observation that the Annada rice variety, commonly used in Andhra Pradesh, India, has a high yield and comparatively low methane emission in that region (Parashar and Bhattacharya, 2002) becomes relevant. Additional research and development appears called for to evolve cultivars that combine significantly low exudations of methane without any reduction in the yield potential appropriate to the location and season for rice.

9.9.1.4 Direct Seeding

Compared to transplanted rice, direct seeding of pre-germinated rice has been reported to reduce methane emissions on account of reduced flooding periods and little soil disturbance. Direct seeding on wet soil and direct seeding on dry soil is seen to reduce methane emissions by 13 and 37 percent, respectively, compared with transplanting eight-day old seedlings (Ko and Kang, 2000). Reduction in methane emissions of 50% due to direct seeding compared to a transplanted crop has been reported (Metra-Corton et al., 2000). Reduction in methane emission of 75% from dry-seeded rice initially flooded at the same time as transplanted rice has been reported (Wang et al., 1999). Dry-seeded rice is given 2 to 3 flood irrigations before permanent flooding. Mineral Nitrogen, applied before flooding, will serve as a source of Nitrous Oxide during the initial drying periods following flooding. Thus, dry seeded rice can be a source of considerable Nitrous Oxide (Freney, 1997). Dry-seeded rice emits very little methane and about 50% more Nitrous Oxide than the transplanted rice. Overall, the total GWP of dry-seeded rice in only about 20% of conventional flooded rice (Pathak, 2011). Direct seeding of rice saves on labour but poses problems of proper land preparation, pest control and early lodging of crop and carries with it a lower yield potential than transplanted rice. Thus, direct seeding as a measure for mitigation of Methane emissions will have a low acceptance level.

9.9.1.5 Systematic Rice Intensification (SRI)

The SRI method of growing rice in irrigated areas is gaining increased acceptance amongst farmers as it saves considerable amounts of irrigation water and also results in higher unit area yields compared to the conventional method of permanently flooded rice. The fluctuations in water table in SRI Rice due to avoidance of flooding should, a priori, result in reduction in Methane emission. Use of mineral nitrogen in SRI Rice is minimal and much less than those of permanently flooded rice. So emission of Nitrous Oxide in SRI Rice will be smaller than that in flooded rice. Use of organic amendments is an integral part of SRI Rice. The organic amendments increase the carbon content of soil (Mandal et al., 2008) and lead to lesser emissions of CO_2, CH_4 and N_2O (Nayak et al., 2007). Enriched soil carbon may, in a later date, result in higher emissions from soil of CO_2 which is less harmful compared to Methane and least harmful compared to Nitrous Oxide in global warming. SRI Rice is expected to lead to a lowering of about 20% of GWP compared to flooded rice (Yue et al., 2005).

9.9.1.6 Minimal Tillage

Emission of methane from flooded rice occurs as diffusion, ebullition and through the rice plant itself. Diffusion of Methane across water surfaces is the least important. Loss of Methane as ebullition (bubbles) is important during land preparation and initial growth, especially in non-clayey soils. As disturbance of soil leads to emission of GHGs from soil and in view of the advances made in weed control and farm machinery, the concept of minimal or nil tillage is being advocated as a mitigation measure against emission of GHGs from soil. However, in case of rice, minimal tillage as in aerobic crops cannot be practiced and considerable amounts of methane from previous crop residues that have been tilled under is released in a very intense way during the tiling stage of rice field preparation. In a standing crop, the indications are that about 80% of Methane emission from flooded rice is by diffusive transport through the aerenchyma system of the rice plant. Nil tillage reduces GHGs from rice by only 10% (Ahmad et al., 2009) vis-a-vis conventional tillage. Direct seeded rice, reduces Methane but increases Nitrous Oxide (Tsurta, 2002), which like Methane can move through the aerenchyma structure of the rice plant (Yu et al., 1996).

From the above, it would appear that (a) adoption of Systematic Rice Intensification methodology and (b) use of (i) new plant types combining low exudation potential with high yields (ii) organic amendments (iii) nitrification inhibitors and (iv) anti-methanogens offer good scope for reducing green house gas emissions from rice. As SRI will result in considerable saving of irrigation water, it will be necessary to see that the saved water is used to increase area of irrigated aerobic crops and not puddled rice.

9.9.2   Mitigation of Nitrous Oxide Emissions from Agricultural Fields

Increasing Nitrous Oxide content of the atmosphere is detrimental to the Ozone layer (Kler et al., 2005). A highly significant source of emission of Nitrous Oxide into the atmosphere from cropped fields is the application of inorganic nitrogenous fertilisers, organic manures, bio-solids and other nitrogenous material in excess of the Nitrogen requirements of crops. The Nitrous Oxide emissions arise from nitrification of mineralised organic Nitrogen in the top soil and denitrification of inorganic Nitrogen in the wet subsoil. The mitigation options are to (i) practice precision-farming in which the rates and timing of nitrogenous fertilisers are so adjusted as to meet the exact nitrogen needs of crops (ii) adopt practices that deliver Nitrogen more efficiently to crops (iii) use fertilisers that release Nitrogen in a slow and controlled manner (iv) properly place fertilisers and (v) use nitrification inhibitors. In view of the perpetual tradeoff between emissions of Methane and Nitrous Oxide in all field mitigation measures, the need to use substances that reduce both Methane and Nitrous Oxide had been pointed out earlier. However, these may be toxic to crops. Use of leguminous, catch or cover crops, in rotation or in mixed cropping, will reduce Nitrous Oxide emissions. Legumes reduce reliance on external nitrogen inputs while catch or cover crops use the extra crop-available nitrogen unused by the previous crop. Legume-derived Nitrogen can, however, be a source of Nitrous Oxide (Rochette and Janzen, 2005). Use of inorganic nitrogenous fertilisers carries the danger of pollution of groundwater resources through leaching. World is fast running out of Liquified Natural Gas (LNG), the raw material needed for manufacture of inorganic nitrogenous fertilisers. In the manufacture of inorganic nitrogenous fertilisers, a lot of Carbon Dioxide also gets released into the atmosphere. It has been reported that cow-dung gas plants have the potential to meet the entire nutritional requirements of two annual crops on all arable lands in India (Firodia, 2004). Thus, biogas slurries need to be increasingly used to mitigate N_2O emissions. The organic route for supplying nitrogen to crops is at present costlier than the inorganic route and efforts must be made to make organic supply of nitrogen to crops cost-compatible. Biogas slurries would lead to an increase in emission of Carbon Dioxide via the soil surface, But CO_2 is only 3% as harmful as N_2O in trapping heat.

9.10 Increasing the Sink Capacity for Carbon Dioxide

The scope for reducing CO_2 emissions from crops and agricultural fields is very limited. This makes the need for increasing the sink capacity for absorption of CO_2 urgent. Increasing (i) the field-occupancy of crops in time and/or space (ii) area under perennial vegetation and (iii) per day unit area yields of crops would constitute additions to sink capacity for CO_2. The ways in which the above can be achieved are detailed below.

9.10.1 Crop Breeding

Higher temperatures, by reducing the field-life duration of crops, will decrease the sink capacity for Carbon Dioxide, The reductions in field crop life are not sufficient to warrant raising of another crop. Breeding crops with higher thermal requirements to ensure no reduction in the field- life duration of crops is one option. Location-wise, the new cultivars that ushered in the Green Revolution had the same field-life duration as the varieties they had replaced. Thus, breeding of crop cultivars with a higher day-degree requirements is a daunting task. Again, cultivars with higher thermal requirements for completion of their life-cycle can only help maintain the status quo of sink capacity for CO_2.

For many crops, at current levels of CO_2, photosynthesis is not source (solar radiation) limited (Borass et al., 2004) but limited by sink (leaves, ears etc.,) capacity (Peet and Kramer, 1980). From the Free-Air Carbon Dioxide Enrichment (FACE) studies (Ainsworth and Long, 2005) and the foregoing discussions, it emerges that despite the envisaged decrease, Solar Radiation would still be supra-optimal for Photosynthesis even at double the present level of CO_2. (Rajvel et al., 2010). So we have to look into other factors for increasing crop yields per unit area per day, such as (a) increase in (i) fraction of interception of solar radiation (ii) radiation use-efficiency and (iii) Harvest Index and (b) reduced respiration rates. There are varietal variations in all the above factors. Thus, the above entities appear to be genetically manipulable.

For any given cultivar, drymatter production is directly proportional to quantum of intercepted solar radiation (Monteith, 1994). The theoretical maximum for interception of solar radiation is 90% (Beadle and Long, 1985). For many crop cultivars the canopy architecture is such that interception of solar radiation is well below the theoretical maximum (Long et al., 2006). By conventional breeding it is possible to substantially improve quantum of capture of solar radiation. Realising a harvest index exceeding the current level of 60% is unlikely (Long et al., 2006). Current levels of radiation use-efficiency of C3 and C4 crops are 70% of theoretical maximum. Genetic manipulation of crop-leaf architecture and RUE and breeding of cultivars for decreased respiration rates (Wilson and Jones, 1982) can significantly increase per day unit area yields and thus, substantially increase sink capacity for CO_2.

9.10.2 Optimal use of Surface Irrigation

With equal crop life-durations under the same environment, a poor yielding cultivar will need as much water as a high yielding cultivar. So the emphasis should be on quantum of water usage per unit area. The argument that reduction in field-life of a crop must lead to a considerably lesser water needs of crops can be contested on the ground that factors leading to crop-life reduction can lead to an equivalent increase in unit area crop water needs. However, elevated CO_2 will reduce crop transpiration. As mentioned earlier under global warming there will be a decrease in evaporative power of air. The combined effect of the two will be a reduction in water need of crops on a unit area basis.

The above gives rise to two options. One is to breed a longer duration cultivar of the same crop to offset the temperature induced reduction in life-duration of the crop and use the savings in irrigation water that would result to proportionately increase the irrigated area of the crops. As mentioned above, breeding cultivars with higher day-degree requirements is a daunting task. So one has to go in for the use of same cultivar and increase its irrigated acreage with the resulting savings in irrigation water. The above will help mitigate or even overcome the yield-reducing influence of higher temperatures by water management as detailed below.

As mentioned earlier a doubling in level of Carbon Dioxide will lead to (a) an increase in (i) temperature of 3 °C and (ii) crop yields of 30% and (b) reduction in (i) transpiration need of around 30% and (ii) evaporative power of air. Doubling of levels of Carbon Dioxide from the 2000 level of 370 ppm is expected to occur by 2100 (IPCC, 1990). The evaporative power of air by 2100 will easily be lower than 3% of the present level. From the foregoing, the corollary is that for every one degree rise in temperature, yield will increase by 10% and water need will go down by 11%. Thus, Prima facie the gross yield for same quantum of water should increase under global warming – the higher the temperature the higher the increase in gross yield for use of same quantum of water. The increase in gross yield for same quantum of water used can be assessed as follows.

As temperature will rise by 3 °C by the end of the century, an assessment covering increases of 1, 2 and 3 °C should do for the present. The base values above which temperatures have to be cumulated varies amongst crops and range from 14 °C for Muskmelon to 2.2 °C for Spinach. For agricultural crops the range is from 4.5 °C for Wheat to 10 °C for Maize. It was thus decided to carry out the assessment for two base temperatures of 4.5 and 10 °C. The percentage reduction in crop life duration for increase in temperature of 1, 2 and 3 °C for crops with base temperatures ranging from 4.5 and 10° C presented by Venkataraman (2004) for mean crop season ambient temperatures of 15, 20, 25 and 30 °C were taken. The reduced durations were divided by 0.89. 0.78 and 0.67 for increase in temperatures of 1, 2 and 3 °C respectively to get the percentage increase in irrigable area for same quantum of water used. From the increased irrigable area figure so obtained, yields were increased by 10%, 20% and 30% for increased temperatures of 1, 2 and 3 °C respectively. Results of computations as detailed above are presented in table 9.1.

Table 9.1 Percentage increase in gross yields for same quantum of water used of crops: A and B with base temperatures of 10 and 4.5 °C respectively

Mean Crop Season Temp	Increase in Mean Air Temperature					
	1 °C		2 °C		3 °C	
	A	B	A	B	A	B
15	102	112	124	129	122	151
20	112	116	127	137	149	163
25	116	117	135	140	161	169
30	117	118	140	143	169	172

The above shows that for crops with low base temperatures growing in colder regime, the increase in gross yield will be higher. For normal crop season temperatures of 20 to 30 °C, the increase will be about 15, 35 and 60 % for increase in mean air temperatures of 1, 2 and 3 °C respectively. The percentage increase in gross yields will apply to all crops and its cultivars as water needs of all crops and cultivars of same field-life duration and growth-rhythm will be the same irrespective of their yield level. Thus, optimal irrigation emerges as the key to increase gross crop yields with available irrigation potential under all likely scenario of global warming. Savings in irrigation water through drip and sprinkler irrigation, switch over to SRI Rice from puddled paddies and minimisation of peak water consumption of crops in summer would be additional mitigating measures.

9.10.3 Social Forestry

Agricultural crops can occupy land for a limited area and finite time only. Surfaces with vegetation like forests, orchards and plantations and perennial crops like bamboo are a major source of global carbon sinks. As mentioned earlier, as opposed to annual crops, such surfaces stand to benefit from the envisaged climate change. Increase in acreage of the above will enhance the sink capacity for absorption of CO_2 emissions. Forests lead the other surfaces in mitigation potential. However, it is difficult to increase forest cover. As things stand at present, deforestation is a significant contributor for emission of Greenhouse Gases. Orchards and plantations require prime land for their culture.

The concept of Social Forestry has, therefore, been mooted as a measure to enhance the sink capacity for CO_2. Social forestry schemes consist of Farm Forestry, Community Forestry, Extension Forestry and Agroforestry, In farm forestry, individual farmers are encouraged to plant trees on their own farmland to provide shade to crops or as wind shelters or for soil conservation. In community forestry, trees are raised on community land and not on private land as in farm forestry. Extension forestry is mostly an urban affair involving planting of trees on the sides of roads, canals and railways and on wastelands. Agroforestry is a sustainable land-use system of growing trees along with agriculture crops on the same piece of land. In practice, the integration between crops and trees is done in a spatial and not temporal manner.

9.10.4 Agro-Forestry

In view of the above, agro-forestry is to be preferred for use as a Carbon sink. However, implementation of agroforestry schemes call for detailed planning relating to selection of trees appropriate to the agroecological zones and adoption of practices for cycling of harvest and regeneration of trees and adoption of measures to mitigate emission of GHGs,. Agroforestry has a mitigation potential equal to 70% of forests vis-a-vis GHGs (Verma, 2006). Area-wise, a sustainable agroforestry system has the potential to offset on an average 15 times its area of deforestation (Dixon, 1995). Agroforestry is viewed as a rehabilitation measure for restoration of degraded forests or of land of low fertility status and production potential. However, proper management of agroforestry systems can increase biomass production 3 to 4 times compared to degraded forests (Maikhuri et al., 2000). Besides sequestration of carbon in form of biomass production, agroforestry is useful in conservation of biodiversity, improving fertility status of soils, enhancing rain water usage and assisting biological control of pests (Pandey, 2007). Use of shade-loving catch or cover crops in agroforestry will enhance land productivity and mitigate Nitrous Oxide emissions from soils.

9.11 Carbon Sequestration

The processes, both natural and anthropogenic, by which CO_2 is removed from the atmosphere or diverted from sources to sinks constitutes Carbon Sequestration. Soil is the bio-geo-chemical interface between the atmosphere, vegetation and hydrosphere and is actively involved in the cycling of atmospheric and terrestrial carbon. Thus, besides vegetation, soils are held to be a significant carbon storage sink. The conversion of carbon dioxide by vegetation biomass constitutes direct carbon sequestration. In soils carbon sequestration occurs due to chemical reactions that convert CO_2 in organic soil carbon to carbonates.

The carbon in the biomass stored as wood in trees can remain sequestered for (i) centuries if used in a non-destructive way such as making of furniture, doors, windows etc., and (ii) for decades if used as fuel. As mentioned earlier burning of biomass from vegetation is a significant source of emission of CO_2. However, if the biomass is converted into a value-added produce like compost and returned to the soil, it is stored as Soil Organic Carbon SOC and the carbon can remain without immediate remission. Soil is also a source of CO_2 and over the years much SOC has been lost due to a variety of reasons. Thus, SOC shows cycles of accumulation and loss and attain a quasi equilibrium state over a period of time (Bhattacharya et al., 2008), which depends on management practices, soil and climate. Loss of SOC is reversible. However, pushing up the SOCs successively to higher quasi equilibrium levels calls for changes in land-use-pattern, vegetation cover and management practices (Bhattacharya et. al. 2008).The capacity of soils for sequestering carbon is uncertain and only about one-tenth of that of vegetation. Current potential for sequestration of carbon in form of SOC by following recommended management practices for agricultural soils is held to be one-fourth to one-third of annual increase in atmospheric CO_2. Smectites improve soil carbon sequestration (Bhattacharya et al., 2008) and in Semi Arid Tropics with Smectite rich soils a nearly fourfold increase in SOC in the last 25 years has been reported (Bhattacharya et al., 2007). In implementing SOC enhancement projects, the trade-off between various competing interests has also to be kept in view.

References

Aggarwal, P.K. 2003. Impact of climate change on Indian Agriculture. Jl. Plant Biol. 30: 189-198.

Aggarwal, P.K. and Mall, R.K. 2002. Climate change and rice yields in diverse agro-environments in India, II. Effects of uncertainties and crop models on impact assessment. Climate Change, 52: 331-343.

Ahmad, S. et al. 2009. Greenhouse gas emission from direct seeding paddy field under different rice tillage systems in central China. Soil and Tillage Research. 106: 54-61.

Ainsworth, E.A. and Long, S.P. 2005. What have we learned from 15 years of free-air CO_2 enrichment (FACE)? A meta-analytic review of the responses of photosynthesis, canopy properties and plant production to rising CO_2. New Physiologist. 165: 351-372.

Ainsworth E. A. et al. 2002. A meta-analysis of elevated CO_2 effects on soybean (Glycine max) physiology, growth and yield. Global Change Biol. 8: 695-709.

Akiyama, H.; Yagi, K. and Yan, X. 2005. Direct N_2O emissions from rice paddy fields: Summary of available data. Global Biogeo-chemical Cycles, 19. GB1005, doi: 10.1029/2004 GB002378.

Allen, L.H. Jr. 1990. Plant responses to rising CO_2 and potential interactions with air pollutants. Jl. Environ. Qual. 19: 15-34.

Allen, L. H., Jr., et al. 1987. Response of vegetation to rising carbon dioxide: Photosynthesis, biomass and seed yield of soybean. Global Biogeochemical Cycles 1: 1-14.

An Yuegai and Li Yuanhua, 2005. Change of evaporation in recent 50 years in Hebei region. Jl. Arid Land Resources and Environ. 19: 159-162 (In Chinese with English summary).

Andrew, N.R. and Hughes, L. 2005. Diversity and assemblage structure of phytophagous Hemiptera along a latitudinal gradient: predicting the potential impacts of climate change. Global Ecological Biogeography. 14: 249-262.

Ashrit, R.G.; Rupa Kumar, K. and Krishna Kumar. K. 2001. ENSO-Monsoon relationships in greenhouse warming scenario. Geophys. Res, Letter. 28: 1727-1730.

Asnani, C.G. 2001. El-Nino of 1997-1998 and Indian Monsoon. Mausam, 52: 57-66.

Attri, S.D. and Rathore, L.S. 2003. Simulation of impact of projected climate change on wheat in India. International Jl. of Climatol. 23: 693-705.

Aulakh, M.S.; Wassmann, R.; Bueno, C. and Rennenberg, 2001. Impact of root exudates of different cultivars and plant development stages of rice (Oryza sativa L.) on methane production in paddy soil. Plant and Soil, 230: 77-86.

Awmack, C.S.; Woodcock, C.M. and Harrington, R. 1997. Climate change may increase vulnerability of aphids to natural enemies. Ecological Entomol. 22: 366-368.

Baker, L.T.; Allen, L.H. and Boote, K. 1990. Growth and yield responses of rice to carbon dioxide. The Jl. of Agric. Sci. 115: 315-320.

Bale, J.S. et al. 2002. Herbivory in global climate change research: direct effects of rising temperatures on insect herbivores. Global Change Biology 8: 1-16.

Bandyopadhyay, A.l.; Bhadra, A.; Raghuwanshi, N.S. and Singh. R. 2009. Temporal trends in estimates of reference evapotranspiration over India. Jl. Hydrologic Eng. 14: 505-515.

Bassirad. H.; Reynolds, J.F.; Virginia, R.A. and Brunelle. M.H. 1997. Growth and root NO^{3-} and PO_4^{3-} uptake capacity of three desert species in response to atmospheric CO_2. Enrichment. Australian Journal of Plant Physiology 24: 353-358.

Bazzaz, F.A. 1990. The response of natural ecosystems in the rising global CO_2. levels. Ann. Rev. Ecol. and Systematics. 21: 167-196.

Beadle, C.L. and Long, S.P. 1985. Photosynthesis- is it limiting to biomass production? Biomass. 8: 19-168.

Bhaskaran, B.; Mitchell, J.F.B.; Levery. J. and Lal, M. 1995. Climate response of Indian sub-continent to doubled CO_2 concentration. International Jl. Climatol. 15: 873-892.

Bhattacharyya, T. et al. 2007. Modeled soil organic carbon stocks and changes in the Indo-Gangetic Plains, India from 1980-2030. Agric. Ecosystem Environ. 122: 84-94.

Bhattacharyya, T. et al. 2008. Soil carbon storage capacity as a tool to prioritize areas for carbon sequestration. Current Sci. 95: 482-494.

Bishnoi. O.P. 1986. Solar radiation and productivity in India. I. Potential productivity. Mausam, 37: 501-506.

Bose, R. 2005. Sliding Tropic of Cancer means harsher climate. Times of India, Pune Edition, 29 June, page 8.

Bronson K.F. and Mosier A.R. 1991. Effect of encapsulated calcium carbide on dinitrogen nitrous oxide, methane and carbon dioxide emissions from flooded rice. Biol. Fert. Soils 11: 116-120.

Brutsaert, W. and Parlange, M.B. 1998. Hydrologic cycle explains the evaporation paradox. Nature, 360: 30.

Casella, E. and Soussana, J.F. 1997. Long term effects of CO_2 enrichment and temperature increase on the carbon balance of a temperate grass sward. Jl. Exptl. Bot. 48: 1309-1321.

Castor, L.L.; Ayers, J.E.; MacNab, A.A. and Krause, R.A. 1975. Computerized forecast system for Stewart's bacterial disease on corn. Plant Dis. Rep. 59: 533-536.

Charlson. R.J. et al. 1992. Climate Forcing by Anthropogenic Aerosols. Science. New Series. 255: 423-430.

Chattopadhyay, N. and Hulme, M. 1997. Evaporation and potential evapotranspiration in India under conditions of recent and future climate change. Agric. and Forest Meteorol. 87: 55-73.

Chen, D. et al. 2005. Comparison of the Thornthwaite method and pan data with the standard Penman-Monteith estimates of reference evapotranspiration in China. Clim. Res. 28: 123-132.

Cicerone, R.J. and Shetter, J.D. 1981. Sources of atmospheric methane: Measurements in rice paddies and a discussion. Jl. of Geophysical Res. 86: 7203-7209.

Cicerone, R.J.; Shetter, J.D. and Delwiche, C.C. 1983. Seasonal variation of methane flux from a California rice paddy. Jl. of Geophysical Res. 88: 11022-11024.

Coakley, S.M.; Scherm, S. and Chakraborty, S. 1999. Climate change and disease management. Ann. Rev. Phytol. 37: 399-426.

Collins, M. 1999. The El-Nino Southern oscillation in the second Hadley centre coupled model and its response to greenhouse warming. Jl. Climate. 13: 1299-1312.

Conway, T.J. et. al. 1988. Atmospheric Carbon Dioxide measurements in the remote global troposphere, 1981-1984. Tell us, 40 B. 81-115.

Cox, P.M. et. al. 2000. Acceleration of global warming due to carbon cycle feedback in a coupled climate model. Nature, 408: 184-187.

Coviella. C. and Trumble, J. 1999. Effects of elevated carbon dioxide on insect-plant interactions. Conserv. Biol. 13: 700-712.

Crisanto, R.E. and Leandro, V.B. 1994. Climate impact assessment for agriculture in the Philippines: Simulation of rice yield under climate change scenarios: Crop modeling study. U.S. Climate Change Division Report, EPA-230-B-94-003. 2-14.

Cure, J. D. and Acock, B. 1986. Crop responses to carbon dioxide doubling: A literature survey. Agric. and Forest Meteorol. 38: 127-145.

Cure, J.D.; Israel, D.W. and Rufty, T.W. 1988. Nitrogen stress effects on growth and seed yield on nodulation of soybean exposed to elevated carbon dioxide. Crop Sci, 28: 671-677.

De, U.S. 2000. ENSO and Monsoon. Vayu Mandal. 30: 1-9.

De Datta, S.K. 1981. Principles and Practices of Rice Production. John Wiley and Sons, New York, USA. 618 pp.

Del Gino, A.D. and Wolf, A.B. 2000. The temperature dependence of liquid water path of low clouds in Southern Great Plains, Jl. Climatol. 13: 3465-3486.

Denier Van Der Gon, H.A.D. et al. 2001. Sulfate-containing amendments to reduce methane emissions from rice fields: Mechanisms, effectiveness and cost. Earth and Environmental Sciences. Mitigation and adaptation strategies for global change. 6: 71-89.

Dixon, R.K. 1995. Agroforestry Systems: Sources or sinks of greenhouse gas? Agrofor. Syst, 31: 99-116.

Dlugokencky, E.J. et. al. 2003. Atmospheric methane levels off: Temporary pause or a new steady state? Geophysical Res, Letters. 30: doi: 10.1029/2003GL018126.

Drake, B.G.; Gonzalez-Meler, M.A. and Long, S.P. 1997. More efficient plants: A consequence of rising atmospheric CO_2. Ann. Rev. Plant Physiol. Plant Molecular Biol. 48: 609-639.

Elizabeth, A.S. and Dina, K, 2006. Global anthropogenic methane and nitrous oxide emissions. The Energy Jl. 27: 33-44.

Fangmeier, A. et al. 1999. Effects on nutrients and grain quality in spring wheat crops grown under elevated CO_2 concentration and stress condition in European, multiple-site experiment ESPACE-Wheat: Eur, Jl. Agron. 10: 215-229.

Firodia, A. 2004. Cows are forever: Methane gas- the answer to oil imports. Times of India. 8-12-04, Pune, India. Page 10.

Food and Agriculture Organisation (FAO). 2008. Climate Related Transboundary Pests and Diseases, Technical Background Document from The Expert Consultation. 59 pp.

Freney, J.R. 1997. Emission of nitrous oxide from soils used for agriculture. Earth and Environmental Sciences: Nutrient Cycling in Agroecosystems, 49: 1-6.

Gadgil, S. and Mishra, S.K. 1995. The Indian climate research programme. Science Plan. 11 pp.

Gadgil, S. et al. 2002. On forecasting the Indian Summer Monsoon: The intriguing season of 2002. Current Science. 83: 394-403.

Gao, G, et al. 2006. Trend of potential evapotranspiration over China during 1956 to 2000. Geophys. Res. 25: 378-387 (In Chinese with English abstract).

Garg, A,; Shukla, P.R.; Kapshe, M. and Deepa, M. 2004. Indian methane and nitrous oxide emissions and mitigation flexibility. Atmospheric Environment, 38: 1965-1977.

Garg, K, N, and Hassan, Q. 2007. Alarming scarcity of water in India. Current Sci. 93: 932-941.

Giorgi, F. et al. 1998. Simulation of regional climate change with global coupled climate models and regional modeling techniques. IPCC. WMO-UNEP Report. Cambridge University Press, U.K. 429-436 pp.

Golubev, V.S. et al. 2001. Evaporation changes over the contiguous United States and the former USSR: A reassessment. Gephys. Res. Lett. 28: 2665-2668.

Goudrian, J. and Unsworth, M.H. 1990. Implications of increasing carbon dioxide and climate change for agricultural productivity and water resources. Amer. Soc. Agron. Spl. Publication No, 23: 111-130.

Hamilton, J.G. et al. 2005. Anthropogenic changes in tropospheric composition increase susceptibility of soybean to insect herbivory. Environ. Entomol. 34: 479-485.

Harrington, R.; Fleming, R. and Woiwood, I.P. 2001. Climate change impacts on insect management and conservation in temperate regions: Can they be predicted? Agric. and Forest Entomol. 3: 153-159.

Harrington R, et al. (2007). Environmental change and the phenology of European aphids. Global Change Biology 13: 1550–64.

Hijmans, R.J; Forbes, G.A. and Walker, T.S. 2000. Estimating the global severity of potato late blight with GIS-linked disease forecast models. Plant Pathol. 49: 697-705.

Hundal. S.S. and Kaur, P. 1996. Climate change and its impact on crop productivity in Punjab. In: Climatic Variability and Agriculture, (Eds) Y.B. Abrol, Sulochana Gadgil and G.B. Pant. Narosa Publishing House, New Delhi, India. 377-393 pp.

Hunter, M.D. 2001. Effects of elevated atmospheric carbon dioxide on insect-plant interactions. Agric. Forest Entomol. 3: 153-159.

Idso, C.D. and Idso, K.E. 2000. Forecasting world food supplies: The impact of rising atmospheric CO_2 concentration. Technology, 7(suppl): 33-56.

Idso, S.B. and Kimball, B.A. 2001. CO_2 enrichment of sour orange trees: 13 years counting. Environmental Experimental Botany. 46: 147-153.

Intergovernmental Panel on Climate Change (IPCC) 1990. Climate Change. The IPCC Scientific Assessment. Houghton, J.T., Jenkins, G.J. and Ephraums, J.J. (Eds). Cambridge University Press, Cambridge, U.K. 410 pp.

IPCC. Climate Change, 1995: The Science of Climatic Change. Summary for Policy Makers. 3-7 pp.

IPCC. Climate Change, 2001. Impacts, adaptation and Vulnerability. Contribution of Working Group II to the third assessment report of the IPCC. J.J. McCarthy et.al. Editors. Cambridge Univ. Press, Cambridge. U.K. 1032 pp.

IPCC. Contribution of Working Group I to Fourth Assessment Report of IPCC. Summary for Policy Makers. 1-18.

Jacobson. M.Z. 2001. Global direct radiative forcing due to multicomponent anthropogenic and natural aerosols. Journal of Geophysical Research. 106: 1551-1568.

Jiang, C. et al. 2006. Methane and nitrous oxide emissions from three paddy rice based cultivation systems in Southwest China. Earth and Environmental Sci. Advances in Atmospheric Sciences, 23: 415-424.

Johnson, H.B.; Polley, H.W. and Mayeux, H.S. 1993. Increasing CO_2 and plant-plant interactions: Effects on natural vegetation. Vegetatio. 104: 157-170.

Karl, T.R. et al. 1991. A new perspective in global warming: Asymmetric increase of day and night temperatures. Geophys. Res. Lett. 18: 2253-2256.

Katenberg, A. et al. 1996. Climate Models: Projection of future climate, In: The Science of Climate Change J.T. Houghton et al. (Eds) Climate Change. Cambridge Univ. Press, Cambridge, U.K. 285-357.

Kaukoranta, T. 1996. Impact of global warming on potato late blight: risk, yield loss and control. Agric. and Food. Sci. in Finland. 5: 311-327.

Keutgen, N. and Chen, K. 2001. Responses of citrus leaf photosynthesis, chlorophyll fluorescence. Macronutrient and carbohydrate contents to elevated CO_2. Jl. of Plant Physiol. 158: 1307-1316.

Khole, M and De, U.S. 1999. Floods and droughts in association with cold and warm ENSO events and related circulation features. Mausam, 50: 355-364.

Khosla, M.D.; Sidhu, B.S. and Benbi, D.K. 2011. Methane emission from rice fields in relation to management of irrigation water. Jl. Environmental Biol. 32: 169-172.

Kimball, B. A. 1983. Carbon dioxide and agricultural yield. An assemblage and analysis of 430 prior observations. Agronomy Jl. 75: 779-788.

Kimball, B.A. and Idso, S.B. 2005. Long-term effects of elevated CO_2 on sour orange trees. In: Plant Responses to Air Pollution and Global Change. Springer, Japan. 73-80 pp.

Kiratani, K. 2007. The impact of global warming and land-use change on the pest status of rice and fruit bugs (Heteropthera) in Japan. Global Change, 13: 1586-1595.

Kler, D.S.; Kaur, N. and Uppal, R.S. 2005. Soil and groundwater pollution by agrochemicals – A Review. Ind. Jl. Environ. and Ecoplan, 7: 285-294.

Ko, J.T. and Kang, H.W. 2000. The effects of cultural practices on methane emission from fields. Nutrient Cycling in Agroecosystems, 58: 311–314.

Kripalani, R.H. and Kulkarni, A. 1997. Climatic impact of El Nino/La Nina on the Indian Monsoon: A new perspective. Weather, 52: 39-46.

Krishan. G. and Saha, S.K. 2008. Potential impact of CO_2 elevation and temperature on plants and its mitigation options. Indian Jl. Environ. & Ecoplan. 15: 327-330.

Krishna Kumar, K.; Kleeman, R.; Cane, M.A. and Rajagopalan, B. 1999a. Epochal Changes in Indian Monsoon precursors. Geophys. Res. Letter, 26: 75-78.

Krishna Kumar. K.; Rajagopalan, B. and Cane, M.A. 1999b. On the weakening relationship between Indian Monsoon and ENSO. Science, 284: 2156-2159.

Kumar. R.K. and Ashrit, R.G. 2001. Regional aspects of global climate change simulations: Validation and assessment of climatic response over Indian Monsoon region to transient increase of greenhouse gases and sulfate aerosols. Mausam, 52: 229-234.

Lal. M. and Singh, R. 2000. Carbon sequestration potential of Indian forests. Environ. Monit. Assess. 60: 315-327.

Lal, M. and Singh, S.K. 2001. Global warming and monsoon climate. Mausam, 52: 245-262.

Lal, M.; Cubasch, U.; Voss, R. and Waszkewitz, 1995. Effect of transient increase in greenhouse gases and sulphate aerosols on monsoon climate. Current Science, 69: 752-763.

Lal. M. et al. 1998. Vulnerability of rice and wheat yields in N.W. India to future changes in climate. Agric. and Forest Meteorol. 89: 101-116.

Lal, M. et al. 1999. Growth and yield responses of Soybean in Madhya Pradesh, India to climate variability and change. Agric. and Forest Meteorol, 93: 53-70.

Lal, M.; Meehl, G.A. and Arblaster, J.M. 2000. Simulation of Indian summer monsoon rainfall and its intra-seasonal variability. Regional Environmental Change. 1: 163-179.

Lal, M. et. al. 2001. Future climate change: Implications for Indian summer monsoon and its variability. Current Science, 81: 1196-1207.

Leakey, A.B. et al. 2006. Photosynthesis, productivity and yield of maize are not affected by open air elevation of CO_2 concentration in the absence of drought. Plant Physiol. 140: 779-790.

Lewis, T. 1997. Thrips as Crop Pests. Commonwealth Agricl. Bur. Cambridge Univ. Press. 740 pp.

Li Chunqiang.; Li Baoguo. and Keqin Hong. 2008. Climate change and its effect on reference evapotranspiration and crop water requirements in Hebei province, China during 1965-1999. Jl. of. Agrometeorol, 10: 261-265.

Liesak, W.; Scnhell, S. and Revsbech, N.P. 2002. Microbiology of flooded rice paddies. FEMS Microbiol. Rev. 24: 625-645.

Liu, B,; Xu, M.; Henderson, M. and Ging, W. 2004. A spatial analysis of pan evaporation trends in China, 1955-2000. Jl. Geophys. Res, 109, D 15102, DOI: 10.1029/2004 JD 004511.

Long. S.P. 1991. Modification of the response of photosynthetic productivity to rising temperature by atmospheric CO_2 concentration: Has its importance been Underestimated? Plant, Cell and Environment. 14: 729-739.

Long, S.P.; Zhu, X.G.; Naidu, S.L. and Ort, D.R. 2006. Can improvement in photosynthesis increase crop yields? Plant, Cell and Environment. 29: 315-330.

Lu, W.F. et al. 2000. Methane emissions and mitigation options in irrigated rice fields in Southeast China. Nutrient Cycling in Agroecosystems. 58: 65-73.

Maikhuri, R. K. et al. 2000. Growth and ecological impacts of traditional agroforestry tree species in Central Himalaya, India. Agrofor. Syst. 48: 257–271.

Majerus, M. and Kearns, P. 1989. Ladybirds, Naturalists' Handbook. Richmond Publishing. Slough.U.K. 103 pp.

Mall. R. K. and Aggarwal, P.K. 2002. Climate change and rice yields in diverse agro-environments in India. Evaluation of impact assessment models. Climate Change 52: 315-330.

Manabe, S. and wetherald, R.T. 1975. The effect of doubling the CO_2 concentration on the climate of a General Circulation Model. Jl. Amospheric Sciences. 32: 3-15.

Manabe, S.; Spleman, M.J. and Stouffer. P.J. 1992. Transient response of a coupled ocean atmospheric model to gradual change of atmospheric CO_2. Part II: "Seasonal Response". Jl. Climatol. 5: 105-126.

Mandal, B. et al. 2008. Potential of double-cropped rice ecology to conserve organic carbon under subtropical climate. Global Change Biology 14: 2139-2151.

Mauney, J.R.; Lewin, K.F.; Hendrey, G.R. and Kimball, B.A. 1992. Growth and yield of cotton exposed to free CO_2 enrichment (FACE). Crit. Rev. Plant Sci. 11: 213-222.

McCarthy, M.P.; Best, M.J. and Betts, R.A. 2010. Climate change in cities due to global warming and urban effects, Geophys. Res, Lett. 37: L09705 doi: 10.1029/2010 G10 2845.

Meehl, G.A. and Washington, W.M. 1993. South Asian summer monsoon variability in a model with doubled atmospheric carbon dioxide concentration, Science, 260: 1101-1104.

Meehl, G.A. and Washington, W.M. 1996. El Nino-like climate change in a model with increased atmospheric CO_2 concentrations. Nature, 382: 56-60.

Metra-Corton, T.M. et al. 2000. Methane emission from irrigated and intensively managed rice fields in Central Luzon (Philippines). Nutrient Cycling in Agroecosystems, 58: 37-53.

Mitchell. J.F.B. and Johns, T.C. 1997. On modification of global warming by sulfate aerosols. Jl. Climate, 10: 245-267.

Mitchell, J.F.B.; Manabe, S.; Tokioka, T. and Meleshko, V. 1990. "Equilibrium Climate Change', in: Houghton, J.T. Jenkins, G. and Ephraums, J.J. (Eds), Climate Change: The IPCC Scientific Assessment. Cambridge University Press, U.K. 131-172 pp.

Mitchell. J.F.B.; Wilson. C.A. and Cunningham, W.M. 1987. On CO_2 sensitivity and model dependence of results, Quarterly. Jl. Roy. Met. Soc. 113: 293-322.

Monteith, J.l. 1994. Validity of the connection between intercepted radiation and biomass. Agric. and Forest Meteorol. 68: 213-220.

Mooley, D.A. and Paolino, D.A. 1989. The response of Indian summer monsoon associated with changes in sea surface temperatures over the eastern south equatorial Pacific. Mausam, 40: 369-380.

Morison, J.I.L. 1987. Intercellular CO_2 concentration and stomatal response. In "Stomatal Function"   Stanford   Univ.   Press. 229-251 pp.

Musser, F.P. and Shelton, A.M. 2005. The influence of post-exposure temperature on the toxicity of insecticides to Ostrinia nubilalis (Lepidoptera: Crambidae). Pest Management Sci. 61: 508-510.

Nayak, D.R., Babu, Y.J., Datta, A. and Adhya, T.K. 2007. Methane Oxidation in an Intensively Cropped Tropical Rice Field Soil under Long-Term Application of Organic Mineral Fertilizers. Jl. of Environmental Quality, 36: 1577-1584.

Neftal, A,; Moor ,E; Oeschger, H. and Stauffer, B. 1985. Evidence from polar ice cores for the increase in atmospheric CO_2 in the past two centuries. Nature, 315: 45-47.

Neelin, J.D. 2009. Rainfall in the climate system: Changes under global warming and challenges for climatic modelers. Essay for Math Awareness Month 2009 "Mathematics and Climate". 6 pp.

Neelin, J.D. et al. 2006. Tropical drying trends in global warming models and observations. Proc. Nat. Acad. Sci. 103: 6110-6115.

Neue, H.U. and Scharpenseel, H.W. 1984. Gaseous products of decomposition of organic matter in submerged soils. In. Organic Matter and Rice. International Rice Res, Inst. Los Banos, Philippines, 311- 328 pp.

Newman, J. A. (2004). Climate change and cereal aphids: the relative effects of increasing CO_2 and temperature on aphid population dynamics. Global Change Biology, 10: 5–15.

Nigam, S. 2003. Monsoon feedback on ENSO evolution. In: Weather and climate Modeling (Eds.V.V.Singh, S, Basu and T.N. Krishnamurthi). New Age Inter Science Publishers. 184-197 pp.

Owensby, C.E.; Coyne, P.E. and Han, J.M. 1993. Biomass production in a tall grass prairie ecosystem exposed to ambient and elevated CO_2. Ecological Applications, 3: 644-659.

Pal. P.K.; Thapliyal, P.K. and Dwivedi, A.K. 2001. Regional climate changes due to double CO_2 simulation by CCM3. Mausam, 52: 221-228.

Pan, Q., Wang, Z. and Quebedeaux, B. 1998. Responses of the apple plant to CO_2 enrichment: changes in photosynthesis, sorbitol, other soluble sugars, and starch. Australian Jl. of Plant Physiology 25: 293-297.

Pandey, D.N. 2007. Multifunctional agroforestry systems in India. Current Science 92 (4): 455-463. ISSN: 0011-3891.

Parashar, D.C. and S. Bhattacharya. 2002. Considerations for methane mitigation from Indian paddy fields. Indian Jl. of Radio & Space Physics. 31: 369-375.

Pathak, H. 2011. Personal Communication.

Patel, H.R.; Patel, V.J. and Vyas Pandey. 2008. Impact assessment of climate change on maize cultivars in middle-Gujarat agroclimatic region using CERES-maize model. Jl. of. Agrometeorol, Special Issue Part-2. 292-295.

Pearcy, R.W. and Bjorkman, O. 1993. Physiological Effects. In: (Ed, Lemon, R,). The response of plants to rising levels of atmospheric carbon dioxide. West view, Boulder, Colorado, U.S.A. 65-106 pp.

Peart, R. M. et al. 1988. Final Report: Impact of Climate Change on Crop Yield in the Southern USA, A simulation study: Institute of Food and Agricultural Sciences, University of Florida, Gainesville, U.S.A.

Peet, M.M. and Kramer, P.J. 1980. Effects of decreasing source-sink ratio in soybean on photosynthesis, photo-respiration, transpiration and yield. Plant, Cell and Environment. 3: 201-206.

Penning de Vries, F.W.T.; Jansen, D.M.; Bergeten, H.F.M. and Bakeman, A. H. 1989. Simulation of eco-physiological processes of growth of several annual crops. Pudoc. Wageningen 271 pp.

Peterson, T.C.; Golubev, V.S. and Groisman, P.Y. 1995. Evaporation losing its strength. Nature, 377: 687-688.

Pruppacher. H.R. and Klett, J.D. Microphysics of Clouds and Precipitation. Kluwer Academic Publishers, The Netherlands D. Reidel. 714 pp.

Rajvel, M. et al. 2010. Effect of diurnal variation of atmospheric and elevated levels of carbon dioxide and photosynthetically active radiation on intercellular concentration and rate of photosynthesis in maize and safflower. Jl. of. Agrometeorol. 12: 1-7.

Rasmusson, E.M. and Carpenter, T.H. 1982. Variations in tropical sea surface temperatures and surface wind fields associated with the southern Oscillation/El Nino. Monthly Weather Rev. 110: 354-384.

Reddy, K.R. and Hodges, H.F. 2000. Climate Change and Global Crop Productivity, CAB International Publishing. Wallingford, Oxon, U.K. 472 pp.

Reddy, K.R. et al. 1998. Interactions of CO_2 enrichment on cotton growth and leaf characteristics. Environmental and Experimental Botany, 39: 117-129.

Reddy, R.M.V.; Krishna Rao. P.V.; Subbiah, S.V. and Venkataraman, L. 2005. Methane efflux studies   from different paddy cultivars grown in South India. Ind. Jl. Environ. & Ecoplan, 10:   581-588.

Reddy, V.R.; Reddy, K.R. and Hodges, H.F. 1995. Carbon dioxide enrichment and temperature effects on canopy cotton photosynthesis, transpiration and water-use efficiency. Field Crops Res. 41: 13-23.

Rochette, P. and Janzen, H.H. 2005. Towards a revised coefficient for estimating N_2O emissions from legumes. Nutrient Cycling in Agroecosystems. 73: 171-179.

Roderik, M.L. and Farquhar, G.D. 2002. The cause of decreased pan evaporation over the past 50 years. Science 298: 1410-1411.

Roderik, M.L. and Farquhar, G.D. 2004. Changes in Australian pan evaporation from 1970-2002. Internatl. Jl. Climatol. 24: 1077-1090.

Roderik, M.L. and Farquhar. G.D. 2005. Changes in New Zealand pan evaporation since 1970s. Internatl. Jl. Climatol. 25: 2031-2039.

Roderik, M.L.; Hobbins, M.T. and Farquhar, G.D. 2009. Pan evaporation trends and the terrestrial water balance: Principles and Observations. Geography Compass 3: 746-760.

Roger, P.A. and Ladha, J.K. 1992. Biological nitrogen fixation in wetland rice fields: Estimation and contribution in nitrogen balance. Plant Soil. 141: 41-55.

Rogers, H.H.; Runion, G.B. and Krupa, S.V. 1994. Plant responses to atmospheric CO_2 enrichment with emphasis on root rhizosphere. Environ. Pollution, 83: 155-189.

Rogers. H.H.; Runion. G.B.; Prior, S.A. and Tobert, H.A. 1999. Responses of plant to elevated atmospheric CO_2: Root growth, mineral nutrition and soil carbon. In: Carbon dioxide and environmental stress. Luo, Y. and Moony, H.A. (Eds). Academic Press, New York, U.S.A., 215-244 pp.

Sadasivam, T.S.; Suryanarayanan, S. and Ramakrishnan, L. 1963. Influence of temperature on rice blast disease. In: The Rice Disease. John Hopkins Press, London. U.K. 163-171 pp.

Samarkoon, A.B. and Gifford, R.M. 1995. Soil water content under plants at high CO_2 concentration and interactions with the direct CO_2 effects: A species comparison, Jl. Biogeography. 22: 193-202.

Santer, B.D. et al. 1995. Towards the detection and attribution of an anthropogenic effect on climate. Climate Dynamics. 12: 77-100.

Sass, R.L. et al. 1992. Methane emission from rice fields: The effect of flood water management. Global Biogeochem. Cycles. 6: 249-262.

Sastry, P.S.N. and Chakravarthy, N.V.K. 1982. Energy mutation indices for wheat crop in India. Agric. Meteorol. 27: 45-48.

Schneider, S.H. 1989. The greenhouse effect: Science and policy. Science, 243: 771-781.

Schutz, H. et al. 1989. A 3-year continuous record on the influence of daytime, season and fertilizer treatment on methane emission rates from an Italian rice paddy. Jl. of. Geophys. Res. 94: 16405-16416.

Setyanto, P. et al. 2004. The effects of rice cultivars on methane emission from irrigated rice field. Indonesian Jl, of Agric. Sci. 5: 20-31.

Shende, R.R. and Chivate, V.R. 2000. Global and diffuse solar radiation exposures at Pune. Mausam, 51: 349-358.

Shin, Y.K. and S.H. Yun. 2000. Varietal differences of methane emission from Korean rice cultivars. Nutrient Cycling in Agroecosystems. 58: 315-319.

Siegenthaler, U. 1990. El Nino and atmospheric CO_2. Nature, 345: 295-296.

Sikka, D.R. 1980. Some aspects of the large scale fluctuations of summer monsoon rainfall over India in relation to fluctuations in the planetary and regional scale circulation parameters. Proc. Ind. Acad. Sci. (Earth and Plant Sci.). 89: 179-195.

Sinha, S.K. 1993. Response of tropical ecosystems to climate change. International Crop Sci. Pub. Crop Sci. Soc. America, Madison, U S A. 282-289 pp.

Sinha, S.K. and Swaminathan, M.S. 1991. Deforestation, climate change and sustainable nutrition security. Climate Change, 16: 33-45.

Song, W. et al., 2010. Distribution and trends in Reference Evapotranspiration in North China plain. Jl. of Irrigation and Drainage Engineering. 136: 240-247.

Srinivasan, J. and Gadgil, S. 2002. Asian brown cloud: fact and fantasy. Current Sci. 83: 586-592.

Stanhill, S. and Cohen, S. 2001. Global dimming: A review of the evidence for widespread and significant reduction in global radiation with discussions on its probable causes and possible agricultural consequences, Agric. and Forest Meteorol. 107: 255-278.

Taylor, K.E. and Penner, J.E. 1994. Response of the climate system to atmospheric aerosols and greenhouse gases. Nature, 369: 734-737.

Tett, S. 1995. Simulation of El Nino Southern Oscillation like variability in a global AOGCM and its response to CO_2 increase. Jl. Climatol. 8: 1473-1502.

Thomas, A. 2000. Spatial and temporal characteristics of potential evapotranspiration trends over China. Internatl. Jl. Climatol. 20: 381-396.

Towprayoon, S. 2003. Mitigation of methane and nitrous oxide emission in the rice field using drainage system. Proc. 3rd International Methane and Nitrous Oxide Mitigation Conference, Beijing, China, 483-496 pp.

Tsuruta, H. 2002. Methane and nitrous oxide emission from rice fields. 17th World Congress of Soil Sci. Paper 2100, 10 pp.

Twoney, S. 1974. Pollution and the planetary albedo. Atmospheric Environment. 8: 1251-1256.

Tyagi, L.; Kumari, B. and Singh, S.N. 2010. Water Management- A tool for methane mitigation from irrigated paddy fields. Sci. of the Total Environment. 408: 1085-1090.

Uprey, D.C. et al. 2006. Carbon dioxide enrichment technologies for crop response studies. Jl. of Scientific and Industrial Res. 65: 859-866.

Venkataraman, S. 1985. Agrometeorological processing of crop water-use data- A study for safflower. Jl. Maharashtra Agric. Univ. 10: 83-85.

Venkataraman, S. 2003. An insight into climate change and future crop prospects in India. Ind. Jl. Environ. & Ecoplan, 7: 483-490.

Venkataraman, S. 2004. On possible reduction in yields of grain crops in future climate. Jl. of Agrometeorol. 6: 213-219.

Verma, K.S. 2006. Harnessing agroforestry potential for mitigating climate change, In: (Eds, Raina, J.N.; Sharma, I.P.; Bawa, R. and Tripathi, D.) Sustainable land resource management in Himalayan region. Publisher, Bishen Singh Mahender Pal Singh, Dehradun, India, 93-106 pp.

Wallin, J.R. and Waggoner, P.E. 1950. The influence of climate on the development and spread of Phytophthora infestans in artificially inoculated potato plots. Plant Dis. Reptr. Suppl. 190: 19-33pp.

Wang, B. and Adachi, K. 2000. Differences among rice cultivars in root exudation, methane oxidation and populations of methanogenic and methanotrophic bacteria in relation to methane emission. *Nutrient Cycling in Agroecosystems.* 58: 349-356.

Wang, B. et al. 1999. Methane emissions from rice fields as affected by organic amendment, water regime, crop establishment and rice cultivar. Environmental Monitoring and Assessment. 57: 213-228.

Wang, Z.Y. et al. 2000. A four-year record of methane emissions from irrigated fields in the Beijing region of China. Nutrient Cycling in Agroecosystems. 58(1-3): 55-63.

Washington, W.M. and Daggupaty, S.M. 1975. Numerical Simulation with NCAR global circulation model of the mean conditions during the Asian Summer monsoon. Monthly Weather Review. 103: 105-114.

Wassmann, R. et al. 1993. First records of a field experiment on fertilizer effects on methane emission from rice field in Hunan Province (PR China), Geophys. Res. Letters. 20: 2071-2074.

Wild, M.; Ohimura, A. and Makowski, K, 2007. Impact of global dimming and brightening on global warming. Geophysical Res. Letters, 34. L.doi:10.1029/2006GLO28031.

Wilson and Jones, J.G. 1982. Effect of selection of dark respiration rate of mature leaves on crop yields of Lolium Perenee cv S 23. Annals of Botany. 49: 313-320.

Xu. X. et al. 2002. Nitrous oxide and methane emissions during rice growth and through rice plants: Effect of dicyandiamide and hydroquinone. Biology and Fertility of Soils, 36: 53-58.

Yagi, K.; Tsuruta, H.; Kanda, K. and Minmi, K. 1996. Effect of water management on methane emission from a Japanese rice field: Automated methane monitoring. Global Biogeochemical Cycles. 10: 255-267.

Yagi, K., Tsuruta, H. and Minami, K. 1997. Possible options for mitigating methane emission from rice cultivation. Earth and Environ. Sci. Nutrient Cycling in Agroecosystems. 49: 213-220.

Yan, X.; Yagi, K.; Akiyama, K. and Akimoto, H. 2005. Statistical analysis of major variables controlling methane emissions from rice fields. Global Change Biol. 11: 1131-1141.

Yao, M.S. and Del Gino, A.D. 1999. Effect of cloud parameterization on the simulation of climate changes in the GISS GCM. Jl. Climatol. 12: 761-779.

Yu, K.W.; Wang, Z.P. and Chen, G.N. 1996. Nitrous oxide and methane transport through rice plant. Earth and Environmental Sci. Biology and Fertility of Soil. 24: 341-343.

Yue, J. et al. 2005. Methane and Nitrous Oxide emissions from rice field and related microorganism in black soil, northeastern China. Earth and Environmental Sciences, Nutrient Cycling in Agroecosystems. 73: 293-301.

Zaitao. P.; Segal. M; Raymond, W.A. and Eugene, S.T. On the potential change in solar radiation over the U.S. due to increase of atmospheric greenhouse gases. Renewable Energy, 29: 1923-1928.

Zhang, A. et al. 2010. Effect of biochar amendment on yield and methane and nitrous oxide emissions from a rice paddy from Tai Lake plain, China. Agriculture, Ecosystems and Environment. 149: 469-475.

Zou, J. et al. 2005. A 3-year field measurement of methane and nitrous oxide emissions from rice paddies in China: Effects of water regime, crop residue and fertiliser application. Global Biogeochemical Cycles, 19, GB2021, doi: 10.1029/2004GB002401.

CHAPTER 10

Agrometeorological Advisory Services

A report on "International Assessment of Agricultural Knowledge, Science and Technology for Development (IAASTD)", carried out at the behest of the World Bank and FAO, had advocated adoption of small scale farming free of use of GM cultivars. Exclusion of GM cultivars has valid merits. However, the reasons advanced for adoption of small scale farming namely that it will lead to greater use-efficiencies of water and nutrients are highly questionable. Further, collection and re-use of rain water, prevention of erosion of top soil by wind and rain and optimal and proper scheduling of irrigation can be done effectively only on a large area basis. Again, control of pests, diseases and weeds is facilitated by mono-cropping over large areas. The obvious assumption in the IAASTD report is that small holdings will be part and parcel of large mono-cropped areas. This is not valid in many a developing country where farm holdings are not only small but are also fragmented. Therefore, the discussions on merits and demerits of small scale and large scale farming can only be of academic interest in many a developing country. Due to the above and the need for ensuring local security of food, oil and pulse crops, the agricultural scenario is one of variegated cropping for any given area and season in many regions of developing countries. As the same weather acts differently on different crops, millions of farmers, with small holdings, both rainfed and irrigated, have to endure every year reduction in potential crop yields on account of weather vagaries.

10.1 Need for Agrometeorological Advisory Services (AAS)

Micro-level planning of cropping systems to suit local climate and cooperative farming to ensure large scale mono-cropping will go a long way in mitigating weather-induced crop losses, However, such planning can only minimise but not totally avoid variegated cropping scenario and will not preclude the subjugation of crops to weather vagaries on an year to year basis, which are recognised as a major crop production risk (George et al., 2005). Occurrences of weather vagaries are outside human control. Agronomic technologies to cope with a variety of weather vagaries such as (a) late start of (i) rains for rainfed crops (ii) season for irrigated crops due to heat waves and late harvest of preceding crop (b) mid-seasonal frosts and droughts (c) weather-induced incidence of pests and diseases and (d) hazardous weather like frosts, high winds, heavy rains etc., are available. For example (i) delay in start of crop season can be countered by using short duration varieties or crops and thicker sowings (ii) effects of frosts can be prevented by resorting to irrigation or lighting up of trash fires (iii) irrigation from (A) harvested rain water can be resorted to for coping with lack of rains and (B) from open or tube wells to cope with heat waves in case irrigated crops. Thus, appropriate forecasts of weather and weather-induced biotic setbacks have significant economic benefits as the cost of pre-facto risk reduction due to weather-effects is much smaller than the post-facto management of the losses (Rathore et al., 2006). Some of the measures like pre-seasonal agronomic corrections, control operations against pests and diseases, supplementary irrigation and pre-poning of crop harvests will be high cost decisions. Therefore, the weather forecasts must not only be timely but must also be very accurate.

10.2 Requirements of Providers of AAS

10.2.1 Crop-Weather Ground Truth

For any given crop, effects of vagaries of any given weather parameter, weather induced attacks of pests and diseases and abiotic stresses like frosts, heat and cold waves, high winds, heavy rains etc., depend on the stage of the crop during which they occur. Thus, for issue of weather-based agronomic advisories, under Agro-meteorological Advisory Services (AAS), it is necessary to have information on the prevalent state and stage of crops, both irrigated and rainfed.

The need for a network of crop-weather stations emerges from the above. Collection of concurrent crop cum weather observations requires use of skilled manpower. Thus, such a net work must consist of minimum possible number of stations that can cover all major rainfed and irrigated crops. Rainfall is extremely variable in time and space. So number of raingauging stations in such a network must be much more than crop-weather stations. For this, it is necessary to delineate homogenous (i) rainfall zones and (ii) climate zones on a crop-wise basis for irrigated crops as detailed earlier. From such zonations, it will be possible to arrive at an optimum and representative network of stations that can provide both crop and weather inputs for an operational AAS. Guidelines for recording observations on phenology, pests, diseases and state of crops have also to be laid down.

10.2.2 Processing of Crop Information and Weather Forecasts

A given anomalous weather situation can have varied effects on crops ranging from physical damage to incidence of pests and diseases. So concurrent crop information and expected weather need to be jointly examined by a panel consisting of agronomist, plant pathologist, plant entomologist and agrometeorologist for framing advisories for field operations that can (i) take advantage of good weather situations and (ii) mitigate effects of bad weather (Hay, 2007). Often weather forecasts cannot be directly used for framing of advisories and some elements of weather would need further processing to obtain derived parameters. In some cases, provision of special codes for transmission of weather data to incorporate parameters not covered routinely for analyses by weather forecasters may be required to be made. Some of the above aspects are mentioned below.

In synoptic meteorology frost is deemed to occur when the screen temperatures reach zero degrees centigrade. However, due to night time radiational cooling of crop canopies, crop-frosts can occur even when screen temperatures are above zero degrees centigrade. The depression of radiation minimum temperature of crops below the screen minimum will vary with places and seasons. Thus, in the winter season, at each station it is necessary to specify on a week-wise/dekad-wise basis the threshold screen minimum temperatures at which crops will be subject to frost. Forecasted wind speeds usually refer to heights of 10 meters. However, it is winds at 1, 2 and 3 meter levels that are of relevance to short, medium and tall crops respectively for application of agrochemical sprays. Thus, similar to minimum temperatures, threshold values of forecasted winds at which protection will be required for short, medium and tall crops have to be specified. Leaf-wetness duration. LWD, is an important parameter in affliction of diseases of crops and as explained above can be deduced from forecasted maximum, minimum and dewpoint temperatures and climatic value of depression of crop radiation minimum temperature below the screen minimum. Thus, each agromet centre must be provided with nomograms for determination of LWD from forecasted temperatures. The Frictional layer near the ground is ignored by the synoptic meteorologist but low level winds in this layer have a great bearing on long-distance dispersal of insects and disease spores. Thus, when warranted the low level wind data has to be specifically culled out. As mentioned earlier crop phenological stage has a great bearing on organisation of protective measures against biotic and non-biotic setbacks to crops. There may be some instances of missing crop information. In such cases the expected phenological crop stage need to be worked out from antecedent crop and/or weather data and from forecasted temperature from pheno-meteorological relationships appropriate to the crop under consideration.

10.2.3 Weather Forecasts

10.2.3.1 Validation of Synoptic Climatology by Weather Forecast Models

The test of a model chosen for synoptic weather forecasting is that it must lead to normal patterns of weather as deduced from climatological considerations when normal values of weather parameters are used in the model. For example a forecast model must give onset date of onset of the Arabian Sea branch of the monsoon as 1st of June over Kerala, India. Similarly models on Cyclogenesis must explain the southward shift of the centers of formation of cyclones in Bay of Bengal, India with the progress of the Indian Southwest monsoon. Thus, if a model cannot explain synoptic climatology it will not be useful for synoptic meteorological weather forecasting.

10.2.3.2 Ensemble and Empirical Models

In the ensemble forecasts, different initial conditions are assumed, to arrive at a mean picture so as to smoothen out the uncertainties and improve forecast reliability. In the empirical methods, for quantitative inputs of weather parameters for the forecast periods in the model, either average values are assumed or values from analogous year(s) are incorporated. Both ensemble forecasts and reference to analogous years call for considerable skill and experience on the part of the forecaster.

10.2.3.3 Present Day Weather Forecasts

Various types of weather forecasts in vogue are classified as Now-Casting, Very Short Range, Short Range, Medium Range, Extended Range and Long Range with period of validity of forecasts up to 2 hours, 12 hours, 72 hours, 10 days, 30 days and one year respectively. As the period of validity of the forecast increases, its reliability decreases. The first three types of forecasts do not give enough time to the farmer to initiate appropriate field measures. Long range forecasts are the least suited for AAS. For example, seasonal, long-range forecasts of monsoonal behaviour do not indicate the temporal distribution of rains by which the forecasted anomaly of rainfall with reference to normal will be realized and hence cannot be used for making pre-seasonal changes in cropping patterns and have only a limited utility of warning of likely delay in start of rainy season in case of a forecast of deficit of total seasonal rainfall. The Medium Range weather forecasts are based on Global Numerical Weather Prediction Models (GNWPMs) and involve solving of complex hydrodynamic equations that give the evolution of atmospheric state from initial conditions. However, small changes from the assumed initial state can affect the forecast state and this limits deterministic predictions of weather to a period of 5 to 10 days only.

10.2.3.4 Suitability of Medium Range Weather Forecasts for AAS

Organisation of measures at the farm level to adapt to or mitigate the effects of adverse weather takes time. Loss of time in framing and communication of forecasts to user-interests is inevitable. So forecasts of weather must have a lead time of at least 5 days and preferably a week. Thus, the Medium Range weather forecasts (MRWFs) offer the highest potential for real-time use as inputs for farming operations (Rathore et al., 2006). The situations in which MRWFs can assist agronomic planning/operations in India have been indicated by Venkataraman (2004). MRWFs suffer from low resolution and there is an urgent need to have high resolution global and meso-scale models to further improve forecasting skills (Rathore et al., 2004) and make output of MRWFs applicable to smaller areas.

10.3 Requirements of User-Interests of AAS

The above discussions relate to the essential requirements for provision of AAS. There is also the question of requirements of user-interests of AAS. The term "User-requirement" is an all pervading one. So for provision of AAS one must catalogue as to what forecast requirements for farming can be catered to (i) with our current scientific capabilities in weather forecasting (ii) inclusion of additional observed factors and (iii) present communication facilities. Some of the requirements of providers and users of AAS are examined below in some detail.

10.3.1 Agricultural Weather Forecasts

The types of forecasts required for critical farming operations have some unique features as detailed below. Land preparation can be done on post-facto receipt of thunder showers. Sowing is never resorted to by farmers without checking adequacy of soil moisture in the seeding zone. Post-sowing agronomic measures to modify soil temperatures and conserve seed zone moisture to ensure proper germination and help seedling establishment are possible. However, clear weather is required for sowing and post-sowing operations. Thus, forecasts of clear weather following a wet spell are crucial. Again, there are areas in which significant thunderstorm activity precedes the arrival of rains associated with well-defined weather systems and the rains once started persist without any let up. In such cases the agronomic strategy is to use pre-seasonal rains for land preparation and resort to dry sowings in anticipation of rain in the next few days. However, dry sown seeds will get baked out in absence of rains. It is prudent to sow on receipt of forecast of impending rains. So forecasts of wet spells become crucial in such areas. Forecasts of dry spells following a wet spell is also required in initiation of control measures against pests and diseases.

10.3.2 Crop-Weather Calendars

Normal features of weather like time of start and duration of crop season and occurrence of rainless periods, frosts and heat waves are taken care of in crop planning based on long local experience.

Agroclimatic analyses can also assist in minimising risks due to variations in temporal march of weather elements. Thus, the emphasis in agromet advisory services is on incidence of abnormal crop and weather situations. Abnormalities in weather situations can be gauged from the readily available data on temporal march of weather elements on a short period and climatological basis. Similarly to assess abnormal crop features one must have normal dates and duration of various crop phases from sowing to harvest.

The week or the Dekad (10 Days) is the accepted time-unit for agrometeorological work. In AAS, it is necessary to know if a given situation relating to crop or weather parameter or both are abnormal and if so in what way. Short period normal values of meteorological parameters are readily available or can be interpolated for a given area. It is necessary to have similar information relating to normal times and duration of phenological crop phases from sowing to harvest with indications against phenological crop stages of their susceptibility to biotic and abiotic stresses and wherever if possible, of weather situations that can trigger such stresses. Depiction of temporal march of such information in a pictorial form is called Crop-Weather Calendar (CWC). Mention of crop-pest and crop-disease calendars have been made earlier.

The CWCs can appraise the farmer of the normal features of crop and weather for a given crop in a given season in his area and help understand the reasons for adoption of cropping and cultural practices recommended in the pre-seasonal and mid-seasonal agromet advisories.

10.3.3   Lab to Land Transfer of Technology

In India the differences in yields of rainfed crops on farmers' fields and nearby research stations have been reported to be increasing with an increase in rainfall (Sivakumar et al., 1983). The above indicates a glaring deficiency in lab to land communication of available agronomic technologies and such communication gaps need to be done away with.

10.4 Communications Infrastructure

Collection of concurrent crop information for (i) the periods and areas of weather forecasts and (ii) meeting of agronomists, crop pathologists and crop entomologists to (a) examine the processed crop-weather information (b) discuss the likely effects of expected weather on standing crops and cultural operations and (c) issue, thereof, advisories for farm operations and (iii) reaching of such advisories to farmers in time for organising requisite field-measures are essential. In view of the likely real-time crop-weather scenario, issue of agromet advisories cannot be centrailsed and consultations amongst experts through video-conferencing will also be required. Keeping in view the literacy level and number of farmers in developing countries, TV and radio services are still the best ways of communicating advisories amongst farmers. Broadcasting of advisories in the local language provides an edge on other means of communication (WMO, 1992; Weiss et al., 2000). Rapid technological advancement in information technology IT and widespread acceptance and use of communication tools emerging thereof, have added the much needed boost for communication of data, information and agromet advisories.

In the last century the communication systems were mostly one-way. In the present century interactive communication systems are being discussed more profoundly. The concept of consultations aimed at bridging the gaps and interest conflicts between providers, users and interpreters of information is applicable to AAS as the farmers would like to verbally interact with nodal agencies or designated field managers issuing agromet advisories. Interactive communication is also needed for getting feedbacks from farmers on the usefulness of agromet advisory services. Interactive communication systems for agricultural advisory services are reported as being adopted commercially by providers and users of agromet advisories in USA, Japan and some European countries. Deliberations, since 2005, of the International Society for Agricultural Meteorology (INSAM) on agrometeorological services have brought out the need for (a) extension intermediaries between providers of AAS and farmers and their training in the mechanics of formation of advisories and (b) interpretation of advisories for organising redemptive farm operations. In developing countries, intermediary personnel have to be provided by governments while in developed countries farmers may be able to employ such personnel.

The communications infrastructure should, therefore, provide for (a) speedy collection and transmission of data, information and advisories (b) quick consultative exchanges amongst specialists and (c) interactive exchanges of intermediaries and farmers.

In recent past a large number of farmers have begun to use mobile phones and can thus use both interactive and non-interactive mode of communications with those engaged in the work of advising them. Farmers can communicate with the web-based systems and can request for advice concerning a newly discovered problem. In the non-interactive mode, mobile-based communication system can be used for receiving the weather forecasts or warning of weather hazards. Short Messaging Service (SMS) for forecasts of weather hazards have now been launched in many countries.

10.5 Limitations, Usefulness and Effectiveness of Agromet Advisories

Adaption and adoption of agronomic technologies to cope with weather vagaries is possible only at the beginning of the season. These include avoiding wasteful and early sowing due to false starts of season, adherence to cut off dates for sowing for various classes of crop varieties and crops, timely land preparation and safe sowing and adoption of agronomic strategies as detailed earlier to cope with late start of season. However, once the crop and its varietal class is sown, the technology and resource-usage gets committed to a particular course of action and the emphasis shifts to coping with mid-seasonal weather vagaries like occurrence of droughts, frosts, high winds and weather-induced incidence of pests and diseases. However, unseasonal rains for clear season crops and persistence of rains beyond normal dates of withdrawal in case of rainfed crops give little scope for agronomic mitigation of weather effects save for prophylactic action against rain-induced incidence of pests and diseases. Late withdrawal of rains in fact reduce yields of irrigated crops through prevalence of cloudiness during crop maturity and unusual breakout of pests and diseases.

Reliability of forecasts and cost-effectiveness of measures against expected weather-induced risks or crop losses will affect the extent of use of agromet advisories, which can be accessed through surveys of farmers (Onyewotu et al., 2003). In developing countries there could be a number of categories of AAS using farmers. Thus, the economic benefits accruing to farmers adopting the tendered advisories under AAS will vary with season, region and crop and for the same crop amongst farmers in different seasons and regions. This is borne out by innumerable published results of surveys, conducted in India on effectiveness of agromet advisories. It is seen that farmers found MRWF and advisories based thereof useful for land preparation/sowing, application of pest and disease control sprays, manures and fertilisers, weed control. Protective irrigation and harvesting and that farmers adopting advisories under AAS got 8 to 10% more yields of both rainfed and irrigated crops than conventional farmers while economic benefits varied from 15 to 65%.

One major outcome from adoption of agromet advisories is in the minimal but effective use of agrochemicals used in the production and production of crops. Reduced use of agrochemicals not only add to the income of farmers through reduced cost of cultivation of crops but also help in the mitigation of global warming. Another significant outcome will be the improved use-efficiencies of rainfall, surface irrigation and ground water.

The expected improvements, in (i) resolution of MRWFs will help applicability of AAS on a small area basis and (ii) interactive verbal communications between farmers and AAS intermediaries will lead to better assessment of the needs of farmers, better provision of products for response farming and active participation of farmers in provision of feedback and developing solutions for specific problems and provision of services thereof.

10.6 Agrometeorological Forecasting of Crop Yields (AFCY)

The need to know sufficiently ahead of harvest, total output of important staple food grain crops, not only on a national basis but on a regional/state wise basis, for central/federal governments and on a district wise basis for state governments has become urgent from points of view of (i) planned, effective and timely organisation of inter and intra-regional movements of food stock from envisaged areas of surplus to those likely to have deficits and (ii) avoiding post-facto situations leading to vulnerability of requisite imports to international market manipulations and high import costs. It is to be expected that such concerns will extend in the near future to agro-industrial, pulse, oil, sugar, fiber and plantation crops and likely to lead to constitution of national crop-weather watch committees on a crop-wise basis. The data inputs for and agronomic advisories emanating from national Agrometeorological Advisory services constitute excellent precursors for crop yield forecasting and AAS infrastructure may be drawn into or form part of national crop weather watches for provision of forecasts of yields of all important crops on a micro-scale. The subject of forecasting yields of rainfed and irrigated crops have been briefly dealt within the chapters on Dry farming Meteorology and Climate Change respectively. Developments in real-time Agrometeorological Forecasting of Crops Yields (AFCY) have gone through the stages of descriptive methods, regression techniques and dynamic Crop-weather simulation. Merits and demerits of the above approaches are briefly mentioned below.

10.6.1 Descriptive Methods

They identify thresholds of one or more variables to classify weather situations relating to groups of significantly different yields. In this no assumption is made of relationship between the chosen variables and yield. The clustering takes account of the inter correlations amongst many weather factors. The inference is that similar weather situations will lead to similar yields. Being non-parametric and requiring no data processing, they lend themselves easy for straight forward applications. Many of the warnings for incidence of pests and diseases and thumb rules for anticipation of crop yields are of the descriptive type. On a local scale, they are as good as the parametric models. However, they are both time and location bound and this limits their applicability for areal assessment of crop yields.

10.6.2 Regression Techniques

In these, crop yield is sought to be related to one or more agrometeorological variable by regression equations. As yields are also influenced by technological inputs, actual yields are normally de-trended with time built in as a regression variable. The data requirement for application of regression techniques is limited and their simplicity lends them suitable for manual calculations. However, there is often no physiological basis between the chosen variables and yield. Similarly there is no physical basis for any connection between the meteorological parameters chosen in the specified time periods. A major disadvantage is the need to develop a series of regression equations for sequential use with progress of the crop season. Thus, they fail badly outside the range of weather variables for which they have been derived due to their over-dependence on too limited a number of factors.

10.6.3 Crop-Weather Simulation

In this, the models take account of the physical and physiological effects of weather factors and the interrelationships between various components on growth, development and yield of crops. All components are described by quantitative variables, which are re-computed at regular intervals and which vary with the level of sophistication of the model. Simulation models are, therefore, more accurate and more versatile. They are, however, complex and require sophisticated analyses of a considerable amount data inputs. They are also specific to management-inputs and cultivars. Despite the above, their usefulness in assessing relative yields in terms of potential yields has been discussed in the sub-chapter relating to impact-assessment on agricultural crop prospects in chapter on climate change and they are now accepted as potential crop forecasting tools. We may now examine the mechanics of issue of Crop Yield forecasts.

10.7   Mechanics of Issue of Crop Yield Forecasts

10.7.1 Date of Forecast

The first point is to determine the earliest date by which a crop yield forecast with acceptable accuracy can be issued. In case of rainfed crops it is obviously the time by which rains withdraw or cease at a location as indicated by the medium range forecast. Many studies have shown that some attributes of crops, like number of panicles per unit area in case of rice, are strongly correlated with yield. Therefore, for irrigated crops, it would be prudent to wait to issue a yield forecast after expression of the yield-oriented crop attribute. For grain crops, many studies have shown a high correlation between yield and crop attributes at ear emergence/anthesis phase. So for grain crops reliable forecasts can be issued after completion of the above phases. The timing for issue of initial yield forecast will vary with crops and can be gauged though statistical analyses of the temporal march of outputs of validated Dynamic Crop Weather Simulation models.

10.7.2 Generation of Future Meteorological Data

The main difference between assessment and forecasting of crop yields by simulation models is the fact that yields forecasting needs an estimation of future meteorological data for the time in the period between framing of the forecast and likely crop harvest. Medium range forecasts of weather parameters can be availed off at the time of framing of the forecast. This can at best cover a period of 10 days. Beyond the above period, several techniques can be considered to be used for the above purpose. In case of rainfed crops the main determining factor in yield estimation in the rain-free maturity period is Reference Crop Evapotranspiration (ET). Inter year variations of ET show little variations. Thus, average values of ET for recent 5 years derived from pan evaporation data and pan coefficients appropriate to the season can be used. Irrigated crops call for adoption of a different approach. One method is to use recorded data of an analogous year (Everingham et al., 2002). These calls for considerable agrometeorological knowledge and experience on the part of the forecaster. The suggestion to use climatological averages is not valid for the simple reason that yields keep improving due to technology and any current yield must relate to weather of as short a period as possible of the recent past but still long enough to retain statistical significance. To meet this challenge, Gommes (1998) suggests the use of Random Weather Generator (RWG), a computer programme which provides time series with the same statistical properties of actual data and training of RWGs with last ten years' data to generate a much longer series to be used in a forecast.

10.7.3 Pests and Diseases

Crop yields are also affected by afflictions of pests and diseases. Agrometeorological prediction of incidence of pests and diseases is possible. By crop sampling techniques it is possible to estimate both the extent and intensity of a pest or disease attack in field crop. Up to the time of issue of forecasts, cognizance can be taken of reports of extent and intensity incidence of a pest and/or a disease. However, due to paucity of available quantitative data on pest and disease incidence, the statistical base for their analyses is weak and simulation models for assessing in quantitative terms reduction from potential yields due to pest and disease afflictions are rare. From forecasted meteorological data one could forecast only likely incidence of a pest or disease but not their effects on crop yields. Thus, effects of large scale attacks of pests and diseases on crop yields have to be assessed separately.

10.7.4 Potential and Relative Yields

As mentioned earlier, while simulation models are likely to fail in predicting unit area yields of a crop in quantitative terms, are cultivar-specific and relate to adopted crop management practices, they are very useful in predicting relative yields i.e., extent of reduction in yield from the potential due to weather vagaries. Potential yield for a given crop cultivar in a given location can be taken as the maximum yield obtained by use of the model in recent past years at the location. The forecasted crop yield should be expressed in terms of percentage fraction of the potential yield. Besides weather, crop yields are influenced by technological inputs as reflected by the presence of a technological trend in any crop data series. Yield forecasts from simulation models need to be and can be weighted for the technology factor by an examination of yields of a crop under consideration in the recent past at or in a nearby location.

10.8   Point, Regional and Gross Yields

Crop yield forecasts issued as detailed above will refer to specific locations and such point data on a micro-scale have to be converted to macro-scale, regional yields, to which all agricultural crop statistics apply. The areal extent to which an observation recorded at a station or an index derived thereof is applicable is always a moot point. As mentioned above, crop yield forecasts from simulation models relate to unit area yields in terms of reduction in potential yield of a crop. The total production in a given region of such a crop depends on the area sown. As mentioned earlier, potential yield of the crop is influenced by "Climatic Fertility" (Bernard, 1992) which often shows marked inter-regional variations. The total production in a given region of such a crop depends on the area sown.

10.9   Satellite Agrometeorology

Areal coverage of earth from space, through suitable satellite-mounted sensors, provides an intermediary meso-scale between micro-scale, point data and macro-scale, regional data. In this geo-stationary satellites help view large areas with feasibility of repeated observations of an area but have a poor resolution. Polar orbiting satellites, which move in an inclined North-South plane at heights varying from 500 to 900 km, on the other hand, give global coverage at fixed intervals of high spatial resolution. They are, therefore, to be preferred for agrometeorological work. The smallest area that can be mapped with satellite sensors varies with satellites.

10.9.1 Sown Acreage of Crops

By sensing the radiation reflected or emitted from bare or cropped surfaces it is possible to locate and identify various crops and crop features. Use of high resolution satellite imagery for crop identification and crop area estimation (Csornai et al., 1990) and for crop differentiation and crop inventory (Dadhwal et al., 2002) is now a well established application. The experiences in India on estimation of area relating to 20 crops in rainy and clear season, over spatial scales ranging from village to region, over a decade, reveals that remote sensing inventory is an operational tool for use in estimation of crop production (Dadhwal et al., 2002). Areal coverage of crops can thus be done on a crop-wise basis and region-wide basis. This constitutes an important step in the complex task of tuning of micro-scale point data of crop yields to derive macro-scale, regional crop yields.

10.9.2 Crop Yield

Initial efforts directly related remote-sensed vegetation indices to crop yield. Recent advances in retrieval from space through remote sensing of biophysical quantities (crop attributes and eco-physiologcal parameters) used in simulation models have opened up many ways of combining appropriate ground-based outputs of simulation models and similar space-based outputs (Moulin et al., 1998). Assimilation of remote sensed agrometeorological parameter through direct incorporation in simulation models is not preferred as it assumes that the remote sensed values are accurate. The emphasis is, therefore, on reconciliation of model and remote-sensed outputs to utilise satellitic observations over large areas at regular intervals. Several techniques (Dadhwal and Bhattacharya, 2004; Chaudhari et al., 2010) have been used for such reconciliation. These include re-initialization and re-parameterization i.e., adjustment respectively of an initial condition and of parameters in the simulation model to obtain agreement between simulated and remote-sensed parameters and forcing in which the value of a crop attribute in the simulation model is corrected by the observed remote sensed value at the same point of time. The most commonly used parameter for such reconciliation is the Normalized Difference Vegetation Index (NDVI) which is positively related to green biomass and Leaf Area Index (LAI) derived from an empirical relationship between field-observed LAI and space-based NDVI at approximately peak vegetative growth. Preference for LAI arises from the fact that LAI is a measure of the crop's ability to intercept solar radiation and helps in understanding the impact of crop management practices (Chen and Cihlar, 1996) - the twin factors regulating crop yield. As updating only one variable may lead to inconsistencies in simulation (Mass, 1988), incorporation of a correction factor in the form of ratio of field LAI to space-based LAI to other crop attributes in the model is recommended (Chaudhari et al., 2010). Such forcing steers the model onto a correct path to simulate yields closer to the actual value.

10.9.3 Gridding

The primary requirement in such an approach is the need to convert all inputs to the same spatial scale. This is facilitated by the technique of "Gridding", in which grid cells of equal area are superposed to the crop distribution map by conical equal area projection. Gridding must be such as to accommodate spatial scales of all factors under consideration and ensure that there is intra-cell homogeneity of soil, weather and crop conditions. Administrative boundaries like districts can be overlaid on the grid map to facilitate issue of forecasts on an administrative unit basis.

10.10   Combining Ground and Satellite Agromet Data

Reconciliation and integrated assimilation of point and space-based data are the next steps. Most suitable model for crop prediction that can be adapted for the above purpose are mechanistic models which require simple inputs (Van Diepen et al., 1989) and which relate physiological growth stage to environmental variables to obtain model output (Chaudhari et al., 2010). Though a remote-sensed parameter can be directly used in a simulation model, it is better to couple it with same data generated from simulation models through processes of forcing, re-initialization, and re-parameterization. Such a procedure helps in developing a corrective relationship between error in some intermediary remote sensed variable and error in final yield. This is necessary for application to cases where yield forecast is required to be made (Dadhiwal et al., 2002). Several studies relating to generation of regional forecasts of yields of many irrigated crops have been carried out. A method combining ground data with satellitic indices like. NDVI is being used by FAO to obtain realistic area-average yields (Gommes. 1993). In this connection, the observations of Csornai et al., (1999) that in the remote sensed crop monitoring and yield forecast models of the Hungarian Agricultural Remote Sensing Program (HARSP), the structure of the model is similar for different crops and it does not depend on the area and the given year's weather is worth noting.

10.10.1   Dates of Sowing

A frequent problem encountered in assessment of regional crop yields is the intra-regional variations in yields caused by variations in dates of sowing of a crop due to a variety of reasons. Such a variation is naturally greatest in case of rainfed crops than irrigated ones. Dates of sowing can be extracted from remote sensed data (i) on spectral emergence as spectral values of near infrared reflectance increase after sowing (ii) from backward extrapolation of the time profiles of NDVI and (iii) on the basis of vegetation fraction cover (*fc*) derived from NDVI. The day with first positive change in NDVI which is greater than soil NDVI is spectral emergence. The time taken for germination is min used from the date of spectral emergence to arrive at the sowing date. Use of remote sensed dates of sowing as input in a simulation model is seen to improve the accuracy of district-level yield forecasts of wheat in India (Seghal et. al., 2002). For dryland crops it stands to reason that rainfall budgeting to derive water requirement satisfaction should commence from the dates of remote-sensed sowing dates in different areas of a region. In case of grain crops, variations in dates sowing dates may or may not result in similar variations of ear emergence/anthesis. It would be prudent to reconcile station and remote-sensed data for the above parameters.

10.10.2   Rainfed Crops

In case of rainfed crops the temporal march of quantum of crop-usable moisture in the root zone is the main yield determining factor. Remote sensing of root zone soil moisture can be a great help in assessment of regional yields of rainfed crops. Both thermal and microwave sensors have been used in surface soil moisture estimation. Satellitic imaging of crops by thermal sensors is not possible during the rainy crop season. Brightness-temperature has an inverse relationship to soil wetness (Sarma and Kaur, 2006). Remote-sensed brightness-temperature data from microwave radiometers can be used to measure soil wetness under different surface roughness and vegetation cover conditions (Ahmed, 1995) and is useful in providing estimates of soil moisture over large areas (Thapliyal et al., 2003). Microwave radiation can penetrate clouds and vegetation cover (Dadhwal and Bhattacharya, 2004) and are to be preferred for soil moisture assessment over rainfed cropped areas. In this, active sensors have to be preferred over passive sensors because of their greater spatial resolution. Wingerton et al., (2003) have reviewed the use of determination of near surface soil moisture by microwave radiometry which can only help assess the soil moisture in the top 10 cm of soil (Dadhwal and Bhattacharya, 2004). However, what is required is quantum of root zone moisture. A remote-sensed assessment of the quantum of root zone moisture available when rains cease and clouds lift-off can help in forecasting yields of rainfed crops. For a crop fully covering the ground the processes of evaporation from surface soil and transpiration from root zone proceed at the same relative rate. Remote sensed soil moisture content in the surface layer has a linear relationship to profile water content (Saha and Mishra, 2006) and relationship between remote sensed surface moisture and profile moisture derived from rainfall budgeting model at a station level at the time of cessation of rains can be used to obtain quantum of profile moisture on a regional scale. Again, remote sensing can help determine crop canopy cum air temperatures. The former is strongly influenced by plant water status (Singh an Kanemasu, 1983) and the difference between canopy and air temperature is measure of moisture stress for the crop. Algebraic addition of such differences from sowing to ear emergence expressed as Stress Degree Days (SDD) (Kingra et al., 2010) can be related to output of quantum of profile moisture from validated models at a station and such a relationship can be used to obtain remote-sensed values of profile moisture on a regional scale.

10.11 FASAL

Polar orbiting earth observation satellite, carrying multi-spectral (visible, near-infrared) sensors, have rendered remote sensed data an important tool for yield modeling. A project "Forecasting Agricultural Output using Space, Agrometeorology and Land-based

Observations" under the Acronym "FASAL" is under implement-tation in India. In future (i) sensors providing hyper-spectral data and with enhanced capabilities of higher spatial resolution (ii) retrieval of additional crop parameters (iii) thermal remote sensing of canopy temperatures and canopy water status (iv) microwave-sensed data of soil moisture (v) improved characterization of crop and its growing environment and (vi) advances in weather forecasting leading to improvements in accuracy, increase in lead-times and catering to smaller areas will lead to (a) better protection of crops from weather vagaries (b) a broadening of the linkage between model-derived and space-sensed data (c) provide additional ways to use point-based crop simulation outputs to account for the spatial and temporal variations in crop yields and (d) more accurate forecasts of yields of many crops, both rainfed and irrigated, on a micro-scale.

10.12 Weather-based Crop Insurance

A risk can be defined as the product of the probability of occurrence of a damaging or hazardous event in either an unforeseeable manner or in a manner that precludes the mitigation of the foreseeable consequences of the occurrence of such an event. Thus, climate and weather constitute major risks in agriculture and the need to manage the above risks has been widely recognised (Sivakumar and Motha, 2007). The starting point for risk-management is risk-avoidance. This can then be followed by risk mitigation and finally by redress of the endured incidence of risk. Now the risks due to climate can be avoided by agroclimatic planning which can (a) identify, on a short-period cum micro-scale basis, endemic areas and periods of abiotic risks (hazardous weather phnonmena) and weather-induced biotic risks (pests and diseases) and (b) help avoid the same by agronomic measures. Some examples are: (i) change in sowing dates and/or choice of suitable cultivars to avoid high temperatures and cyclonic weather during crop maturity (ii) specifying cut-off dates for sowing long, medium and short duration varieties on a crop-wise, region-wide basis both under rainfed and irrigated conditions to ensure optimal yield and avoid inclement weather during crop maturity (iii) change of sowing dates and mixed cropping to avoid endemic pests and diseases (iv) arranging for irrigation or trash fires during endemic frost periods and (v) assess, for dryland crops, occurrence of times and period of (vi) incidence of soil moisture stress and (vii) availability and quantum of surface run-off to help construct on-farm-reservoirs for collection and re-use of surface runoff from rainfall harvesting for re-use to combat drought situations.

As detailed above, agronomic advisories based on medium range forecasts of expected weather and prevalent crop situation can help a farmer to (i) mitigate the consequences of bad weather (ii) maximise the effects of good weather and (iii) ensure, minimal and yet effective protection to crops from pests and diseases. However, early warning systems and weather forecasts, which can provide timely information to farmers to take necessary measures to totally annul the impact of weather anomalies and weather-induced set-backs, are still not perfect and are in various stages of development for realizing the ultimate goal. As mentioned earlier, because of (i) the small size of land holdings and need for regional self-sufficiency in food security of food, pulse and oil crops the crop composition for any area given area and season presents a variegated picture. For a given place and season abrupt inter-year variations (i) in incidence, spread and intensity of crop pests and diseases and (ii) from the normal short period weather features regularly occur. Due to variations amongst crops and their growth stages in susceptibility to biotic and abiotic stressed to realised weather, millions of farmers annually run the risk of reduction in crop losses. The setbacks to crop production and hardships to farmers caused by weather vagaries and pests and diseases are widespread and regular and visible enough to attract both public and governmental concern for provision of quick relief to the affected farmers.

10.12.1 Coverage

As risks are covered by insurance, the concept of crop insurance is topical and relevant. Crop insurance will free the farmers of the anxiety in awaiting timely and adequate relief payments for compensating set backs in crop production. Crop insurance will also free (i) the state governments, in a federal structure, of submission of reports of crop losses due to weather anomalies and pests and diseases and (ii) the central government of the responsibility of verification of crop-loss reports submitted. To be effective crop-insurance cover needs to cover all crops and must be provided to all farmers.

10.12.2 Fixation of Premia

In any insurance business, the premium depends on the risk level - the higher perceived risk, the higher the premium. In crop insurance also, from the business point of view, premium will have to be higher for areas with unstable crop production. Higher crop insurance premia would be an unbearable burden for farmers in areas of low yields of high inter-annual variability of low value crops. The governments have, therefore, to pay either the entire crop-insurance premia or at least the differences in actual premia paid compared to the lowest one for a given crop, region and season. Payments of entire premia as well as subsidisation of premia under crop insurance for farmers cannot be treated as subsidizing uneconomic crop production, even under existing WTO regulations.

10.12.2.1 Irrigated Crops

Agroclimatic methodologies, such as dynamic crop-weather simulation, can be used to analyse past series of meteorological data to assess, for many irrigated crops and for any given area and season, the extent of inter-annual variability in crop production. The above approach is justifiable since the differences between cultivars to weather parameters are one of degree rather than that of type. For any given irrigated crop, such an agroclimatic evaluation, even if relative across (i) regions in a season and (ii) years in a place would help in arriving at properly weighted premia in tune with the risk level.

10.12.2.2 Dryland Crops

Compared to irrigated crops determination of the variability in production of dryland crops is more complex. For this delineation of homogeneous rainfall zones in the dry-farming tract as outlined in the chapter on Dryfarming Meteorology is a must. The probability data on short-period minimum assured rainfall amounts can be used to delineate rainfall zones in the dry-farming tract with intra-zonal homogeneity with reference to start and end of agriculturally significant rains and duration of crop-life period (Venkataraman, 2001). For representative stations, for each of the district in each zone, the duration of crop-life period can be worked out at probabilities of 30, 50 and 70%. In dryland agriculture 50% probability is an acceptable risk level. In considering inter-annual variability in dryland crop production, one can consider the extent of reduction in crop life at 30% probability vis-a-vis that at 50% as a measure of inter-annual variability in dryland crop production at a place.

10.13 Factual Assessment of Crop Losses

10.13.1 Dryland Crops

Recently use of satellite imageries to assist evaluation of claims under crop insurance has been made. It is a commendable idea, as it will provide an independent check on claims of yield losses of crops. Cloud cover renders satellite imaging of Kharif crops very difficult and irregular. However, it is possible to arrange for satellite imaging as soon as the cloud cover lifts and during periods of crop harvests. Satellite imageries of Normalised Differential Vegetative Index, NDVI, can help determine extent of crop cover while remote sensed differences in crop canopy and air temperatures along with NDVI can give a measure of the status of soil moisture stress of crops (Moran et al., 1994) when rains cease. Satellite-sensed moisture stress when calibrated against ground-truth can give a measure of moisture available for crop use when rains cease. This in turn can be used to assess the likely soil moisture stress at harvest time and hence of the extent of reduction in yields vis-a-vis a non-stressed crop. Thus, satellite imaging can be used in case of insurance claims of rainfed crops also.

10.13.2 Clear Season Crops

In case of clear season crops, periodic satellite-imaging of temperatures and cloudiness would be required to assist settlement of crop insurance claims. Ground-truth relating to crop weather features will be required to validate the satellite-sensed imageries and have to be arranged to be provided through joint efforts of meteorological and agricultural departments.

References

Ahmed, N.U. 1995. Estimating soil moisture from 6 GHz dual polarization and/or satellite derived vegetation index. Int. Jl. Remote Sensing. 16: 687-708.

Bernard, E.A. 1992. L'intensification de la production agricole par l'agrometeorologie. Agrometeorology working papers series No. 1, FAO, Rome 35 pp.

Chaudhari, K.N.; Tripathy, R. and Patel, N.K. 2010. Spatial wheat yield prediction using crop simulation model, GIS remote sensing and ground observed data. Jl. of Agrometeorol. 12: 174-180.

Chen, J.M. and Cihlar, J. 1996. Retrieving leaf area index of boreal conifer forest using Landsat TM images. Remote Sensing Environ. 55: 153-162.

Csornai, G.,; Dalia, O.; Farkasfaly, J. and Nádor, G., 1990. Crop Inventory Studies Using Landsat Data on Large Area in Hungary, Applications of Remote Sensing in Agriculture, Butterworths, 159-165 pp.

Csornai, G. et al., 1999. Crop Monitoring by Remote Sensing. Paper presented at the FIG Commission 3rd Annual Meeting and Seminar. Budapest, Hungary. 21-23 October.

Dadhwal, V.K. and Bhattacharya, B.K. 2004. Remote sensing in agrometeorology. Jl. of Agrometeorol. 6, Spl. Issue: 144-152.

Dadhwal, V.K.; Singh, R.P.; Dutta, S. and Parihar, J.S. 2002. Remote sensing based crop inventory: A review of Indian experience. Tropical Ecol. 43: 107-122.

Everingham, Y.L. et. al., 2002. Enhanced risk management and decision-making capability across the sugarcane industry value chain based on seasonal climate forecasts. Agric. Systems, 74: 459-477.

George, D.A. et al., 2005. Surveying and assessing climate risk to improve farm business management. Extension Farming Systems Jl. 1: 71-77.

Gommes, R. 1993. The integration of remote sensing and agrometeorology in FAO. Adv. Remote Sensing. 2: 133-140.

Gommes, R. 1998. Agrometeorological crop yield forecasting methods. Proc. International Conf. on Agricultural Statistics, Washington, 18-20 March. Eds. Theresa Holland and Marcel P.R. Van Den Broecke. International. Statistical Inst. Voorburg, The Netherlands. 133-141 pp.

Hay, J. 2007. Extreme weather and climate events and farming risks. In Managing Weather and climate Risks. Eds. M.V.K. Sivakumar and R, Motha. Springer, Berlin, Heidelberg. 1-19 pp.

Kingra, P.K. et al., 2010. Prediction of grain yield of wheat using canopy temperature based indices. Jl. of Agrometeorol. 12: 58-60.

Mass, S.J. 1988. Using satellite data to improve model estimates of crop yield. Agron. Jl. 80: 655-662.

Moulin, S.; Bondeau, A. and Delecolle. R. 1998. Combining agricultural crop models and satellite observations: from field to regional scales. Internatl. Jl. Remote Sensing, 19: 1021-1036.

Moran, M.S.; Clarke, T.R.; Inoue, Y. and Vidal, A. 1994. Estimating crop water deficit using the relation between surface-air temperature and spectral vegetation index. Remote Sens. Environ. 49: 246-263.

Onyewotu, L.O.Z. et al., 2003. Reclamation of desertified farmlands and consequences of its farmers in semiarid northern Nigeria: A case study of Yambawa rehabilitation scheme. Arid Land Research and Management 17: 85-101.

Rathore, L.S.; Singh. K.K. and Gupta, A. 2004. National centre for medium range weather forecasting: activities, current status and future plans. Jl. of Agrometeorology, 6: Spl. Issue 258-264.

Rathore, L.S., Maini. P and Kaushik, S. 2006. Impact assessment of the agro-meteorological advisory service of the National Centre for Medium Range Weather Forecast (NCMRWF). Available on the INSAM website (www.agrometeorology.org) under Accounts of operational agrometeorology.

Saha, R. and Mishra, V.K. 2006. Estimation of profile moisture status from surface moisture in hilly slopes of Meghalaya. Jl. of Agrometeorol. 8: 81-86.

Sarma, A.A.L.N. and Lakshmi Kumar, T.V. 2006. Studies on agroclimatic elements and soil wetness using MSMR data, Jl. of Agrometeorol. 8: 19-27.

Saghel, V.K.; Rajak, D.R.; Chaudhary, K.N. and Dadhwal, V.K. 2002. Improved regional yield prediction by crop growth monitoring system using remote sensing derived crop phenology. International Archives of the Photogrammetry, Remote Sens. & Spatial Inf. Sci. 34 Pt.7, 329-334.

Singh, P. and Kanemasu, E.T. 1983. Leaf and canopy temperatures of pearl millet genotypes under irrigation and non irrigation conditions. Agron. Jl. 75: 477-501.

Sivakumar, M.V.K.; Singh, P. and Williams, J.H. 1983. In Alfisols in the Semi-Arid Tropics: A consultants' workshop, 1-3 December, 1983, ICRISAT Centre, India. 15-30 pp.

Sivakumar, M.V.K. and Motha, R. 2007. Managing weather and climate risks in agriculture. Springer, Berlin, Heidelberg. 503pp.

Thapliyal, P.K.; Rao, B.M.; Pal, P.K. and Das, H.P. 2003. Potential of IRS-P4 microwave radiometer data for soil moisture estimation over India. Mausam, 54: 277-286.

Van Diepen, C.; Rappold, C.; Wolf, J. and van Keulen, H. 1989. WOFOST: A simulation model of crop production. Soil Use Management, 5: 16-24.

Venkataraman, S. 2001. A simple and rational agroclimatic method for rainfall zonations in dryland areas. Indian Jl. Ecoplan. & Environ. 5: 431-436.

Venkataraman, S. 2004. Climatic Characterisation of Crop Productivity and Input-Needs for Agrometeorological Services. Jl. of Agrometeorol. 6: 98-105.

Weiss, A.; Van Crowder, L. and Bernardi, M. 2000. Communicating agrometeorological information to farming communities. Agric. and Forest Meteorol. 103: 185-196.

Wigneron, J.P. et al. 2003. Retrieving near surface soil moisture from microwave radiometric observations: Current status and future plans. Remote Sens. Environ. 85: 489-506.

WMO, 1992. La radio rurale et la diffusion des informations agrometeorologiques. Proceedings of an International Workshop. WMO, Geneva.

CHAPTER 11

Agricultural Renewal and Sustainability

Sustainability of agricultural crop production can be defined (Venkataraman, 2003 a) as the "planned identification, location and efficient use of input-resources and climate to meet the agricultural crop needs of the expected increase in population without quantitative depletion of natural resources and qualitative degradation of the edaphic, aerial and hydro environments". By the above definition in many regions of the world agricultural development has entered the downward slope of non-sustainability. The above situation is due to anthropogenic agricultural activities in total disregard of the effects of climatic elements on the (i) production and protection of crops and (ii) water and nutrition requirements of crops as detailed below. Irrigation of crops in excess of their water needs have led to salinisation of soils. Over-use of fungicides, pesticides and weedicides have polluted soils and water resources. Injudicious use of inorganic nitrogenous fertilisers have resulted in degradation of soils, contamination of water resources and injection of Nitrous Oxide into the atmosphere and hence has been the most harmful. Deforestation in general has accelerated soil erosion and in catchment areas of rivers have led to reductions in capacity and lives of reservoirs on account of soil sedimentation. It has also contributed to an increase in CO_2 concentration in air. Last, but not the least, is the unbridled petroleum-based energy usage leading to ingestion of greenhouse gases into the atmosphere resulting in global warming, which will more severely punish repetition of past agricultural follies of neglect of climatic effects and misuse and overuse of production resources. In light of the above we may consider the most essentially required actions and procedural formalities required thereof for agricultural renewal and sustainability. While considering agricultural renewal, one should keep in mind the need for ensuring food security for the unorganised farm laborers who constitute the economically weakest segment of the population.

11.1 Cooperative Farming

As indicated earlier, climate oriented mono-cropping of large areas have definite advantages by way of (i) collection and re-use of rain water (ii) prevention of erosion of top soil by wind and rain (iii) optimal and effective scheduling of irrigation (iv) minimal and yet effective control of pests, diseases and weeds and (v) better provision and utilization of agrometeorological advisories and hence should be preferred. In many a developing country farm holdings are small and fragmented and precludes utilisation of above benefits. Again, fragmented holdings lead to variegated cropping for any given area and season. As different crops react differently to any given situation of weather, millions of farmers with small holdings suffer crop losses annually from vagaries of weather. In the present circumstances the obvious solution lies in ensuring that small holdings become part and parcel of large mono-cropped areas to derive all the benefits that will accrue from large scale farming. Formation of farming cooperatives is the only available solution to realise the above aim. Towards this end, the governments should act as facilitators by (i) educating the farmers of the need for cooperative farming and (ii) training farmers in cooperative crop management. Reports appearing in the media from time to time speak of efforts of individuals in successfully and profitably organising cooperative farming in several areas and this is a good augury.

11.2   Optimal and Conjunctive use of Water Resources

11.2.1 Surface Irrigation

11.2.1.1 Optimal Use

As detailed in chapter 9, the effects reduction in crop yields due to a shortening of field-life duration of crops on account of higher temperatures resulting from elevated levels of CO_2 will be more than compensated for all crop cultivars in all areas and seasons due to (i) reduction in transpiration and (ii) increase in yield under elevated levels of CO_2 and (iii) decrease in evaporative power of air. The above gives rise to the possibility of using the resulting savings in irrigation water to proportionately increase the irrigated area of the crops with available existing irrigation potential.

11.2.1.2 Savings

The need for avoiding non-crop usage of water needs no emphasis. One such move will be to switch over from flooded rice to SRI Rice, which will also mitigate global warming by lesser emission of methane. Since SRI Rice yields substantially more than flooded rice the saved water should be used to increase acreage of irrigated aerobic crops. Minimisation of irrigated acreage of standing crops in summer and use of drip and sprinkler irrigation are also valuable tools in saving irrigation water.

11.2.2 Groundwater

From the point of view of reservoir operations, water can be released only at uniform intervals. Thus, surface irrigation is available to farmers only at specific pre-determined intervals. Groundwater thus constitutes a readily utilisable source of irrigation. In a warmer atmosphere, extreme weather events like frosts, heat waves and droughts are slated to increase. Thus, as mentioned earlier, to improve water use efficiency of groundwater it is necessary to eschew its use as a sole source of irrigation and use it strategically in providing supplementary irrigation to combat (i) failure of the beneficial pre-seasonal thunderstorms (ii) mid-seasonal droughts, cold and heat waves and (iii) in starting sowing operations in time in case of delay in start of rains and/or availability of surface irrigation. Cooperative farming will provide a de-facto social ownership of groundwater reserves and help (i) achieve the above aims by limiting groundwater usage to average annual replenishment and (ii) in augmentation of groundwater reserves through organisation of measures to improve infiltration of rainfall.

11.2.3 Rainfall

In view of the expected increase in (i) occurrence of moderate and heavy rains at the expense of lighter ones and (ii) variability in time and space of rainfall, construction of on-farm-reservoirs to collect surface runoff from rains for re-use to save dryland crops from mid-seasonal droughts is urgently called for. But this can only be facilitated by cooperative land ownership.

11.3 Organic Cropping

Liquefied Natural Gas, LNG, is the raw material for manufacture of the inorganic nitrogenous fertilisers, during which large amounts of CO_2 are emitted. World is likely to run out of LNG in 3 to 4 decades. Denitrification of nitrogenous fertilisers produces Dinitogen Oxide, which has the potential to destroy the Ozone layer. (Kler et al., 2005). The nitrogen requirements of crops can be met from organic sources. For example Firodia (2004) reports that slurries from cow dung gas plants have the potential to meet the entire nutritional requirements of two annual crops raised on all arable lands in India. Organic manures will emit CO_2, which is only 3% as harmful as N_2O emitted from use of inorganic nitrogen fertilisers in global warming. Thus, replacement of inorganic fertilisers by organic manures will lessen the Global Warming Potential from agricultural practices. The Agrochemicals used for protection of crops from pests, diseases and weeds are of inorganic origin and often not biodegradable and hence ecocidal. GM cropping is not suited and hence cannot be recommended for controlling pests, diseases and weeds. Non-chemical methods like (i) mixed cropping (ii) biological control though introduction of exotic natural parasites and predators of pests and diseases (iii) manual and/or mechanical weeding and (iv) avoidance of pests and diseases through change in sowing dates and/or cultivars together with initiation of control operations only when warranted as per agromet advisories will go a long way in reducing the quantum of use of biocides. Organic insecticides and fungicides of plant origin (Pandey et al., 2004; Dwivedi and Shekhawat, 2005) are (i) very effective in controlling pests and diseases (ii) biodegradable and (iii) eco-friendly. They can, therefore, easily replace the currently used inorganic ones.

Thus, a switch over to organic farming commends itself. However, at present the cost of providing mineral nutrition and protection to crops by organic means is more costly than the inorganic route. Efforts, therefore, need to be made to lower cost of provision of nutrients and crop protection by organic means. With increasing scarcity and hence cost of raw materials needed to produce inorganic biocides, organic farming will become cost competitive.

11.4   Integrated Management of Pests and Diseases (IMPD)

One feature that is not yet sufficiently appreciated is that elevated CO_2 and higher temperatures due to global warming will singly and/or in combination act in a synergic or opposing way on development of weeds, pests and diseases. This will, in any given area and season, lead to an alteration in (i) the composition of and times and duration pests, diseases and weeds and (ii) the times and duration of occurrence and infectivity of pests and diseases. The resultant of the combined effects cannot yet be foreseen on a real-time basis. In case of weeds action can be initiated after perusal of their incidence. However, for pests and diseases prophylactic action must be initiated when their population level is low i.e., as soon as they are sighted.

Catches in spore traps and insect traps can assist in forewarning of diseases and pests. In this the concepts of Biofix, Economic Threshold Levels, ETL of insects and Critical Disease Levels. CDL of diseases as detailed in chapter 7 come into play. Catches in insect and spore traps are seen to be useful in anticipating outbreak of pests and diseases. Thus, any IPDM programme must lay down ETL and CDL criteria for important crop pests and diseases and include a network of observing and reporting spore and insect trap centers. Also, the current information on areal and seasonal incidence of pests, diseases and weeds need to be updated keeping the expected, future climatic scenario.

11.5   Climatic Crop Planning

Life duration and developmental rhythm of crop cultivars are weather-controlled. This leads to wide area variations in the unit area economic yields of any given cultivar even under irrigation in a season and inter-seasonal variations at a location. The term "Climatic Fertility" has been used (Bernard, 1992) to stress the direct link between climatic variability in weather parameters and variations in yield potentials of crop cultivars. Thus, climate also constitutes a crop production input. In irrigated relay cropping, because of agroecological reasons, even for a given crop its rotational crops cannot be the same in all regions. Differences in composition of relay crops will only marginally affect the total crop water needs. However, for maximization of crop outturns, it is necessary to use crops and its varieties suited to local climate. For this the crops must be so chosen that their phasic weather requirements mesh with the temporal march of the concerned weather parameters at a location and season. Agroclimatic methodologies can be used to determine for a given region and season (i) the maximal field-occupancy time possible under irrigated relay cropping and (ii) type of crop cultivars and their sequencing that would ensure maximal yield of each of the crops.

Agroclimatic determination of times and duration of crop-life period for a given distribution of rainfall can be used to gauge, on a location-specific basis, the frequencies of occurrence of specified times and duration of crop-life periods and thus aid in the planning of the dryland cropping system most suited to local rainfall climatology.

11.5.1 Specialised Production of Agro-Industrial Crops

Growing of staple food crops whenever and wherever possible, irrespective of climatic suitability for economic production, is justifiable from the point of view of immediate and local availability of food crops and subsidisation of such production is well called for. However, ubiquitous growing of industrial crops under a pricing scheme to ensure profitability to the farmers in areas and seasons of low productivity of the crop carries the risks of (a) widening the inter-regional disparity in income of farmers growing such a crop (b) differences in profits of farmers supplying their produce to mills/factories at specified dates through staggered sowings/plantings (c) non-exportability of surpluses and (d) danger of import dumping of produce at a price lesser than what the consumer pays domestically in absence of such imports. The solution to the above problems posed by the current scenario of industrial cropping lies in going in for regionalised. Specialised and economic production of specific industrial crops for realising the climatic-fertility potential. For this the entire spectrum of national requirements, production and pricing of all agro-based industrial crops need to be gone into, keeping in perspective national needs, export capabilities, protection from import-dumping, profit security to farmers and fairness to consumers.

11.6   Evolving of New Varieties

As new land cannot be opened up for crop culture, increase in unit area of crop yields in unit time becomes imperative. For this evolving of new varieties is a must. The need for cultivars with (i) physiological senescence during maturity and with ability to reduce water uptake during soil moisture stress as in case of traditional varieties (Venkataraman, 1981) for use in dryland agriculture (ii) leaf-architecture that enables penetration of radiation to ensure that the lower most leaves also photosynthesise optimally and (iii) higher thermal requirements for completion of life cycle to cope with warmer temperatures has been mentioned earlier.

As mentioned in the discussions in the chapter on climate change, while there is little scope for increasing Harvest Index from current level of 60% by conventional breeding, it is possible to evolve cultivars with canopy leaf structure to intercept the theoretical maximum value of 90% (Beadle and Long, 1985). As production of drymatter is linearly related to intercepted solar radiation (Monteith, 1994), the above will lead to highly significant increases in crop yields even with no improvement in their Harvest Indices. The consensus is that the chances of significantly increasing RUE by breeding in the near future seem slight. However, selection and/or breeding of crops for decreased respiration rates holds promise of substantial increase in crop yields (Wilson and Jones, 1982). A combination of all the above mentioned improvements in yield-influencing traits of cultivars will, one can dare to guess-estimate a doubling of per day of unit area crop yields.

11.7   GM Cultivars Free Farming

Regarding GM crops none of the cultivars have been shown to out-yield their non-GM ones. The two main reasons advocated for use of GM crops are their resistance to weeds and pests. The advocacy of GM crops for control of weeds is unnecessary and of pests is premature. The feasibility of pests and weeds developing resistance to the GM crop toxin is ever present. The fears of contamination of soil environment and non-GM cultivars are genuine. What mars GM technology the most is the inability of the farmers to reproduce the seeds, retaining their original genuineness for sowing their next crop. Thus, the farmers are at the mercy of seed companies and are easy targets for touts pandering non-genuine F2 material. If the cost of procurement of seeds for next sowing is taken into account, the net income from cultivation of GM cultivars will be much less than those involving normal cultivars. In view of the ban in many countries on import of GM crop produce, the interests of Indian agricultural export trade lies in GM free cropping. Therefore, the nation would do well to stick to non-GM crops and seek their genetic improvement for higher production by the conventional means. Genetic modification as a means of breeding new varieties can be ruled out on account of the negative aspects and deleterious effects of GM cropping as mentioned above. Thus, for the present it is necessary to use conventional crop breeding techniques.

11.8   Pricing Support for Crop Produce

Even under irrigation unit area yields of crops show inter-regional and inter-seasonal variations. One and same Minimum Support Price (MSP) for any given crop will widen the inter-regional disparities in income of farmers. So MSP should be fixed on a regional and seasonal basis for any given crop.

11.9   Weather Management of Crops

Delays in start of Kharif and Rabi cropping seasons, due respectively to rainfall and temperature vagaries, are not infrequent at a place. Issue of seasonal weather forecasts can then be used to implement contingency cropping plans. Once a crop is sown, resources and technology get committed to a particular course of action. Even then, action to mitigate the effects of unfavourable weather and maximise the effects of favourable weather can be taken on the basis of agrometeorological advisories based on medium range weather forecasts and condition of standing crops. Timely and effective prophylactic action to ward off pests and disease attacks can be taken on the basis of agromet advisories. To achieve this (i) every village, headed by a trained climate manager, must become an IT-connected information centre and (ii) medium range weather forecasts must be improved for their issue on a district-wise basis. However, reports in India of increase in differences in yields of areas vis-a-vis yields at nearby research stations with increasing rainfall is a disturbing feature which emphasises the need for better communications in transfer of knowledge from lab to land and between users and providers of agromet services.

11.10   Weather-based Crop Insurance

Despite climate-based planning and weather-based management of crops, as detailed above, crop losses due to weather vagaries affecting large number of farmers cannot be ruled out. The only way to provide well- targeted, timely and adequate relief to crop growers suffering from weather vagaries is through crop insurance. In this the Governments should bear the entire crop insurance premium in case of marginal, small farmers growing staple crops, subsidise the same in case of farmers with large holdings and encourage horticultural farmers and orchard owners to go in for crop insurance.

11.11   Food Security

Measures for agricultural sustainability as suggested above will help in taking care of the plight of land-owning farmers. However, the end aim of all agricultural development is to ensure food security, as envisaged by Swaminathan and Medrano (2004), to the weakest section of the populace to which the landless agricultural labourers belong. For this the labourers have to be provided with non-farm employment in schemes such as Rural Employment guarantee schemes in the non-crop season so as to avoid creation of labour shortage for crop operations. Such schemes should be carried out as food for work programmes to create permanent assets that would help future crop production such as construction of weather and pests proof food storage structures and on-farm-reservoirs for collection and re-use of surface runoff from rains for alleviation of crop droughts. Other avenues would be participation in post-harvest preservation and processing of produce, value addition to crop biomass and crop-produce based small-scale industries. Agroforestry schemes can provide year round gainful employment to farm labourers during their off days from farm work.

11.12   Concerted Action

Thus, (a) cooperative and organic farming (b) optimal and conjunctive use of water resources (c) breeding of cultivars for (i) maximal interception of solar radiation (ii) increased leaf-photosynthetic rates and (iii) reduced respiration rates (d) climate-based (i) crop planning and (ii) fixation of support prices for crop produce (e) better management of crops through agromet advisory services (f) weather-based and government subsidised crop insurance (g) agronomic and genetic manipulations to improve use-efficiencies of production inputs (h) well organised network system for integrated management of pests and diseases (i) investments in rural areas for creating infrastructure for storage and movement of crop produce and (j) creation of permanent assets like on-farm-reservoirs, water storage bodies etc., though provision of non-farm employment to landless labourers in the off-season in food for work programmes hold the key for combating the challenges of global warming and achieving agricultural sustainability.

References

Beadle, C.L. and Long, S.P. 1985. Photosynthesis- is it limiting to biomass production? Biomass. 8: 119-168.

Bernard, E.A. 1992. L'intensification de la production agricole par l'agrometeorologie. FAO Agrometeorology Working Papers Series No.1. FAO, Rome. 35 pp.

Dwivedi, S.C. and Shekhawat. N.B. 2005. Studies on efficiency of five botanical extracts as pupicidal agent against Trogoderma granarium (Everts). Ind. Jl. Ecoplan & Environ.10: 31-34.

Firodia, A. 2004. Cows are forever: Methane gas- the answer to oil imports. Times of India. 8-12-04, Pune, India. Page 10.

Kler, D.S.; Kaur, N. and Uppal, R.S. 2005. Soil and groundwater pollution by agrochemicals- A Review Ind. Jl. Environ & Ecoplan. 7: 285-294.

Monteith, J.L. 1994. Validity of the connection between intercepted radiation and biomass. Agric. and Forest Meteorol. 68: 213-220.

Pandey, U.K.; Pandey, V. and Singh, P. 2004. Response of some plant origin insecticides against Spodoptera litura (Tobacco caterpillar) infesting some food plants. Jl. Curr. Sci. 5: 737-739.

Swaminathan, M.S. and Medarano. P. 2004. Towards a Food Secure India: A call for policy initiatives and public action. Publication M.S. Swaminathan Research Foundation and world Food Production Programme of FAO, I to IV pp.

Venkataraman, S. 1981. Lysimetric observations on moisture accretion for and use by M-35-1 jowar at Solapur. Jl. Maharashtra Agric. Universities. 6: 36-40.

Venkataraman, S. 2003. Impact of climatic factors on sustainability of crop production with environmental conservation in developing countries.

Wilson, D. and Jones, J.G. 1982. Effect of selection of dark respiration rate of mature leaves on crop yields of Lolium-Perenne cv S23. Annals of Botany. 49: 313-320.

Index

A

B